HEAT AND THERMODYNAMICS

PHYSICS IN ACTION

HEAT AND THERMODYNAMICS

Elizabeth H. Oakes

CHELSEA HOUSE
An Infobase Learning Company

To my family, where I first encountered the laws that govern the natural world

Heat and Thermodynamics
Copyright © 2012 by Infobase Learning

Chelsea House
An imprint of Infobase Learning
132 West 31st Street
New York NY 10001

Library of Congress Cataloging-in-Publication Data

Oakes, Elizabeth H., 1964-
  Heat and thermodynamics / Elizabeth H. Oakes.
      p. cm. — (Physics in action)
  Includes bibliographical references and index.
  ISBN 978-1-61753-100-6
  1. Heat. 2. Thermodynamics. I. Title.
  QC254.2.O24 2012
  536—dc23
                              2011044776

Chelsea House books are available at special discounts when purchased in bulk quantities for businesses, associations, institutions, or sales promotions. Please call our Special Sales Department in New York at (212) 967-8800 or (800) 322-8755.

You can find Chelsea House on the World Wide Web at
http://www.infobaselearning.com

Excerpts included herewith have been reprinted by permission of the copyright holders; the author has made every effort to contact copyright holders. The publishers will be glad to rectify, in future editions, any errors or omissions brought to their notice.

Text design by James Scotto-Lavino
Composition by EJB Publishing Services
Illustrations by Sholto Ainslie
Photo research by Elizabeth H. Oakes
Cover printed by Yurchak Printing, Landisville, Pa.
Book printed and bound by Yurchak Printing, Landisville, Pa.
Date printed: July 2012
Printed in the United States of America   ⁻

10 9 8 7 6 5 4 3 2 1

This book is printed on acid-free paper.

CONTENTS

ACKNOWLEDGMENTS

Huge gratitude goes to Cai Nowicki, without whose help this book would have been impossible. Thanks, also, to my family for their patience and encouragement, and to Frank K. Darmstadt, my editor of many years, without whose faith and support I would never have had this opportunity.

OVERVIEW

The study of heat and thermodynamics is really the study of what happens to energy in the universe. Every single thing, living and nonliving, is made of energy. In other words, matter is energy. In order to understand how matter behaves, we must understand how energy works, and energy, also known as atomic motion, is heat. Heat can do many things—it can cook food, melt ice, warm a cold room, and burn down a house. It can transform what it touches, move from one location to another, and excite atoms into greater activity or calm them down by leaving. The study of heat's behavior teaches us about the movement of energy; the study of thermodynamics teaches us about the movement of energy in and out of systems. Without an understanding of the effects of heat and the laws of thermodynamics, the modern study of physics would be impossible.

In chapter 1, we discuss the basics of heat, temperature, and expansion of materials. This chapter also introduces the scales and units used to measure temperature, and how temperature and heat are related to each other. Finally, this chapter covers the behavior of matter as it is exposed to changing temperatures.

Chapter 2 goes in-depth on the kinetic theory of heat and discusses the idea of heat in motion in an ideal gas. This chapter also explains how kinetic theory is connected to such events as evaporation, and finally discusses the idea of the mean free path of atomic motion.

The focus of chapter 3 is the study of heat transfer. This includes all methods by which heat can be transferred from one object to another: convection, conduction, radiation, and mass transfer. This chapter again includes a discussion of how heat transfer affects the behavior of matter and also how it relates to living organisms.

Chapter 4 discusses the idea of phase changes, the transitions between states of matter. It covers the different types of phase changes that exist and the types of matter between which they occur. Next, the heat of fusion and its effects on heat transfer between objects is analyzed. The idea of enthalpy and how it relates to phase changes is introduced.

In chapter 5, we introduce the idea of thermodynamics, starting with the history and discovery of the field and how those discoveries affect science today. This chapter also discusses the many branches of the study of thermodynamics, and then finally shows how thermodynamics relates to mechanical work.

The laws of thermodynamics are discussed in chapter 6. There are actually four such laws, beginning with the relatively new zeroth law. This chapter analyzes all four, and explains how they affect the behavior of matter in our universe. The relation of these laws to the current study of physics is also briefly discussed.

Without an understanding of these seemingly simple concepts, a full comprehension of physics cannot exist. Heat and the laws of thermodynamics control many aspects of the attributes of matter, and change the way we interact with the universe. So many occurrences of everyday life relate to heat, from choosing the clothes we should wear, to cooking our food, and even to the mechanics of the vehicles we drive.

1

Temperature, Heat, and Expansion

Heat and temperature, while not the same phenomenon, are inextricably linked to each other. An understanding of these seemingly simple concepts is vital to a full comprehension of most topics in physics. This is due in great part to the relation of heat with atomic movement and expansion of materials. This chapter will discuss the physics of heat and temperature, show how they affect the behavior of different materials, and lay the foundation for a wider discussion of heat exchange, kinetic theory, and thermodynamics in the rest of this text.

HEAT

As we know, all matter in the universe is composed of smaller units called atoms. These atoms, and the molecules that they combine into, are in constant motion, even within objects that seem to be holding still. The speed at which these atoms are traveling—one way that the energy these atoms contain is given

1

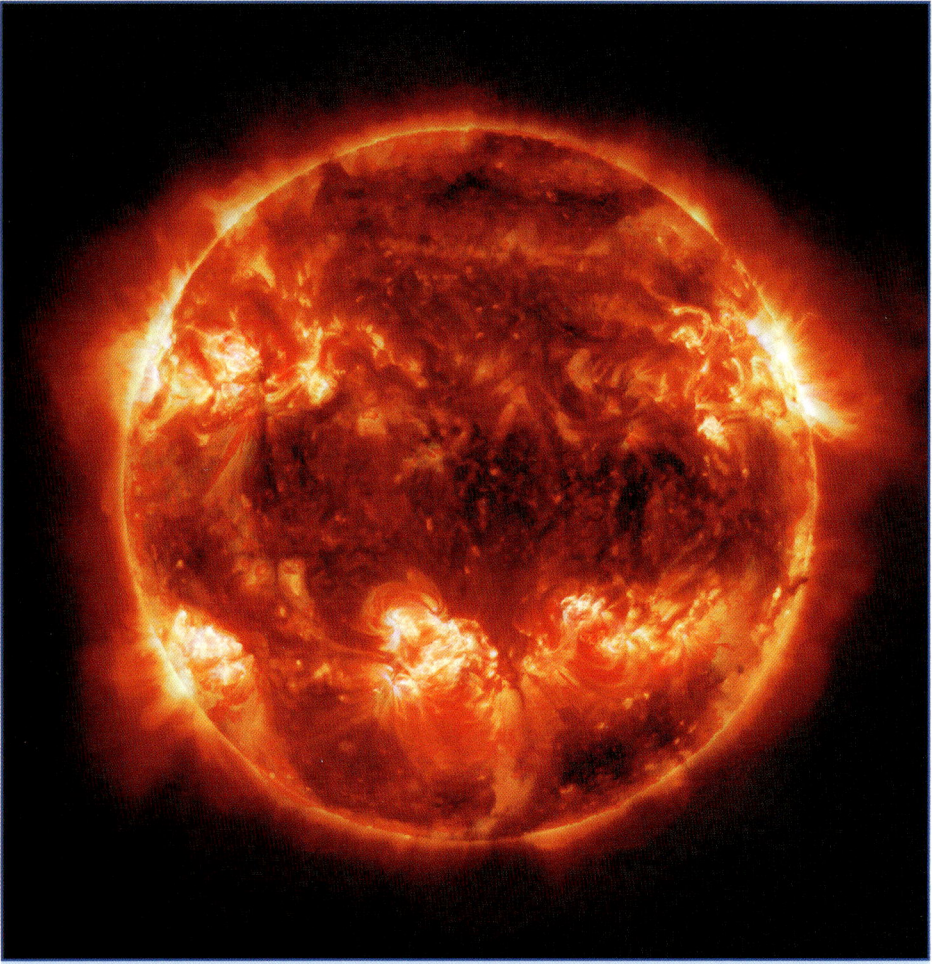

The Sun, the first source of heat energy (NASAJESA)

off by the atoms—translates through our senses into what we call **heat**. However, only lateral movement, movement along a straight or curved line, is related to heat; atoms wiggling or rotating do not impact the temperature of the object. Picture a fishing pole, where the amount of line that is run out represents the amount of heat contained in that object. If the hook (representing the atoms) is pulled downward or wound upward, the amount of heat changes; the hook can jiggle freely in other ways that do not affect the amount of line run off the fishing pole, just as an atom can wiggle in place without affecting the amount of heat in an object.

A further discussion of this kinetic theory of heat will take place in chapter 2, but the basics include the effect of molecular speed on the heat and state of the matter being observed. Materials with faster atoms will feel hotter than materials with relatively slower atoms. Solid material is made up of relatively slow-moving atoms; as they increase in speed, and therefore in the heat energy they contain, the material becomes a liquid, and then after this a gas or even a plasma. Phase changes—these shifts between solid, liquid, and gas—will be discussed further in chapter 5.

Increasing the heat of that object means increasing the total **internal energy** of that object, which is the sum of all types of energy that the molecules in the object contain. While not all of the energy of the atoms in an object is expressed as heat, the energy represented as heat is one part of the internal energy of a piece of matter. The heat energy of an object can be measured using units of either **calories** or **joules**. One calorie is equivalent to 4.186 J. It is important to remember, however, that a hot object does not contain the heat; instead, the object contains energy that becomes heat as it is transferred from one object to another. For example, the heat felt on your skin when you touch a hot object is not a property intrinsic to the object but instead your body's perception of the energy as it is transferred from the hot object to your hand. Heat is transferred from one object to another through the collision of atoms, which will result in a transfer of kinetic energy from the first object to the second. Heat will always travel from an area of higher movement to an area of lower movement. Touching a hot surface such as the side of a cooking pot moves the energy from the pot to your skin, while touching a cold object such as an ice cube moves energy from your skin to the ice. So once again, the feeling of heat or cold is actually a transfer of energy from one object to another.

To continue with the previous example, while your hand and the object are touching and exchanging heat energy, they are considered to be in a state of **thermal contact**. Thermal contact is defined as any situation where energy can be transferred between two objects in the form of heat. While in this example they are physically in contact with each other, thermal contact does not necessarily require the two objects to be touching. This depends on the type of heat transfer the objects are undergoing, a concept that will be discussed further in chapter 3. Once they reach the same temperature, the two objects are said to be in a state of **thermal equilibrium**. Once these objects reach this point of thermal equilibrium, just like in any other equilibrium in physics, the energy on each side remains constant and is no longer being moved from one to the other. The two objects will then remain at that same temperature unless energy is being added to the **system** from another source. An example of this phenomenon is the use of a thermometer to measure temperature. The thermometer and its **environment**, such as the air, a glass of water, or the mouth of a hospital patient, will reach equilibrium, and the thermometer will then display the temperature at that equilibrium, which is also the temperature of the other object.

Another basic concept in heat and temperature is the idea of *specific heat*. The **specific heat** of a substance is a measurement not of the actual heat of a substance, but of its ability to absorb and hold onto **thermal** energy when the temperature of its environment changes. **Specific heat capacity** is the amount of heat energy that must be applied to a gram of the substance to increase the temperature by one degree Celsius. Different materials will have different specific heats. Water, for example, has a very high specific heat of one calorie per gram, which is higher than most other common materials. It takes a lot of heat to raise the temperature of a small amount of water, and it also takes a long time for the water to release that heat. This property is what makes water such a useful coolant and also an effective heating agent in such situations as radiators and hot water bottles. It also affects the climate in many parts of the world, since the slow

Calculating Specific Heat

Let's say we're doing an experiment to determine the specific heat of some unknown metal. The metal weight has a mass of 15 g, and can be assumed to be at the room temperature of 25°C. Our calorimeter contains 100 g of water, heated to 60°C. We place the weight into the calorimeter with the water and wait for the system to reach equilibrium. The final temperature of the water after the system reaches equilibrium is 59.5°C. Using our known values, we can calculate Q:

$$Q = 1 \text{ cal/g°C} * 100 \text{ g} * (60°C - 59.5°C)$$
$$Q = 1 \text{ cal/g°C} * 100 \text{ g} * (0.5°C)$$
$$Q = 100 \text{ cal/°C} * 0.5°C$$
$$Q = 50 \text{ cal}$$

Now, the only value left unknown in our equation is the specific heat of our metal. Plugging in the values we determined:

$$50 \text{ cal} = c * 15 \text{ g} * (59.5°C - 25°C)$$
$$50 \text{ cal} = c * 15 \text{ g} * (34.5°C)$$
$$50 \text{ cal} = c * 517.5 \text{ g°C}$$
$$c = 0.096 \text{ cal/g°C}$$

So the specific heat of our mystery metal is 0.096 cal/g°C!

release of heat from the ocean moderates the weather in coastal areas. The air surrounding the ocean loses its heat much faster, due to having a lower specific heat, and so the climate in many coastal areas would be cooler if it were not for the heat energy being released from the ocean waters.

The specific heat of a substance can be calculated using the mathematical equation $Q = c * m * \Delta T$. In this equation, Q is the heat energy that is added to the system, measured in calories (cal) or joules (J); c represents the specific heat, given with the units cal/g°C or J/g°C; m is the mass of the test object; and ΔT is the change in temperature during the experiment, measured in °C, and is simply the absolute value of the final temperature of the system subtracted from the initial temperature. Researchers have used this formula to determine the specific heats of various materials through experimental testing. This testing involves taking a known quantity of some material at a known temperature and placing it in a calorimeter, a device that is used to measure energy change in a system by containing the heat within itself, along with a known quantity of water heated to a known temperature different from that of the test material. The water and the test substance will come to thermal equilibrium, and this equilibrium temperature is the final temperature that can now be used to calculate ΔT. Since the heat energy gained by the test material is equal to the heat energy lost by the water, and the specific heat of water is already known to be 1 cal/g°C, the value of Q can be calculated using the specific heat equation.

TEMPERATURE

Temperature, unlike heat, is not an actual physical property of an object but instead a measurement of the object's heat relative to a designated point. Temperature measures the average speed of the individual atoms in an object, while heat is the total thermal energy of that object. Therefore, it is important to remember that an object such as a spark can have a very high temperature while transmitting very little heat to another object with which it comes in contact, such as when you use your fingers to pinch out a match or a candle flame without being burnt. This is because while the individual atoms are traveling very quickly and the temperature of the candlewick is very high, there are only a few of them, so they do not have much total energy to impart to the contacted object and your skin remains unburned. This is also relevant to the previously mentioned idea of thermal equilibrium being reached by passing the heat energy down the temperature gradient: The heat always passes from the hotter object to the cooler one, not necessarily from the one with the highest internal energy. In the example of your fingers pinching out a candle flame, your hand has many more molecules and therefore more total energy than the candlewick, but the energy passes from the candle to your hand nonetheless.

As mentioned, temperature is a measurement of the heat energy in an object relative to another designated point. The points that heat energy is related to are known as temperature scales. Temperature is measured using three main scales: Fahrenheit (°F), Celsius (°C), and Kelvin (K). The Fahrenheit scale that most people in the United States are familiar with is calibrated with the freezing temperature of water placed at 32° and the boiling point at 212°, assuming standard atmospheric pressure is present. However, most of the world and most scientific calculations use the Celsius scale, which calibrates with water freezing at 0° and boiling at 100°, again at standard atmospheric pressure. While the Celsius scale is easier to imagine in your head and to use to compare materials to each other, since it works as a percentage of the boiling point of water, Fahrenheit is more exact, by having smaller degree intervals.

Known for his contributions to the laws of thermodynamics and the discovery of absolute zero, Lord Thomson Kelvin was knighted by Queen Victoria for his work on the transatlantic telegraph project. (Dibner Library of the History of Science and Technology, Smithsonian Institution Libraries)

The third scale, Kelvin, differs from Fahrenheit and Celsius in that it uses a completely different starting point for determining relative temperature: **absolute zero**. This makes a good starting point for measuring temperature, as by definition there can be no temperatures that are colder than absolute zero. In fact, nothing can actually reach absolute zero; the reasons behind this barrier will be discussed further in chapter 7. Therefore, while both the Fahrenheit and Celsius scales reach into the negative, generally, to represent temperatures colder than the freezing point of water, the Kelvin scale does not have any negative temperatures. The Kelvin scale also does not use the degree unit that both the Fahrenheit and Celsius scales use. However, the steps between points on the scale are the same as on the Celsius scale, so a difference of 30°C is the same as a difference of 30 K.

Converting temperatures between the Celsius and Kelvin scales is fairly straightforward. Absolute zero is measured as –273.15°C and as 0 K, so to convert a temperature from K to °C is merely a case of adding 273.15 and the degree unit. Converting from °C to K obviously then only requires subtracting 273.15 and removing the degree symbol. Converting between Celsius and Fahrenheit is also straightforward, though it does require a more complicated formula to account for the differences in calibration

Kelvin's Many Contributions

Considered to be the foremost scientist of his time, the Irish physicist William Thomson Kelvin (1824–1907) originated the absolute scale of temperature now named after him, the Kelvin scale, and the concept of absolute zero, the temperature at which substances hold no thermal energy. He is also credited with developing the first law of thermodynamics, known as the conservation law of energy. His long-term collaboration with fellow English physicist James Prescott Joule (1818–89) resulted in the postulation of the Joule-Thomson effect, which describes the change in temperature that accompanies expansion of a gas without production of work or transfer of heat.

points and size of the units. The formula for converting from Fahrenheit to Celsius is (°F − 32°) * 5/9 = °C, which makes the converse formula from Celsius to Fahrenheit °F = 9/5 * °C + 32°.

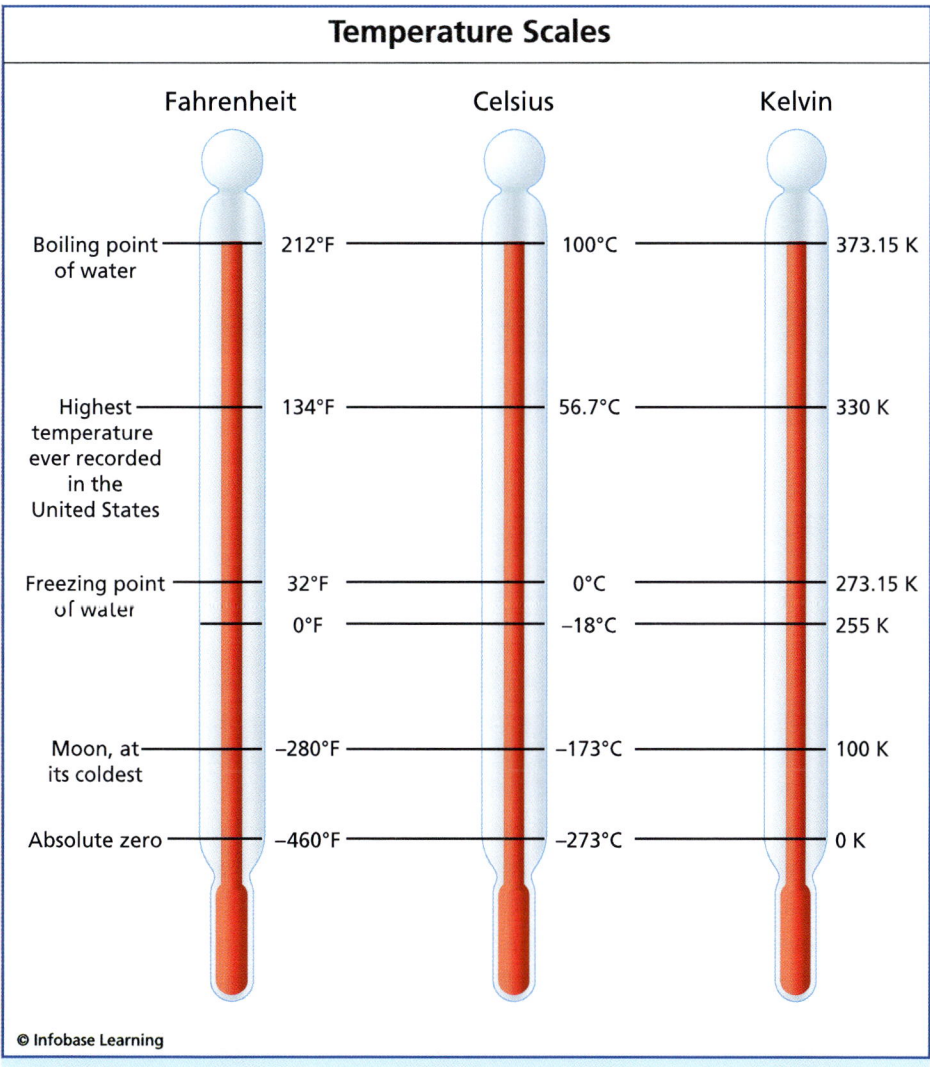

Temperature Scales

Fahrenheit	Celsius	Kelvin
Boiling point of water — 212°F	100°C	373.15 K
Highest temperature ever recorded in the United States — 134°F	56.7°C	330 K
Freezing point of water — 32°F	0°C	273.15 K
0°F	−18°C	255 K
Moon, at its coldest — −280°F	−173°C	100 K
Absolute zero — −460°F	−273°C	0 K

© Infobase Learning

This diagram compares the three temperature scales across the same points. Fahrenheit and Celsius use the same calibration points whereas Kelvin is based on absolute zero. The Kelvin temperature scale starts at absolute zero (-273.15°C); centigrade starts at the freezing point of water.

Comparing Temperatures

Kelvin: T(K) = T(C) + 273.15
Centigrade: T(C)=(5/9)*[T(F)-32F]

At very high temperatures, there is little difference between kelvin and centigrade temperatures. For example, the average surface temperature of the Sun is 5,800 K that is 5,527 C. The two temperatures are nearly the same. On the other hand, Fahrenheit is much different. The Sun at 5,800 K is around 10,000°F!

T(K) = 5,800 Kelvin = 1.8 x 5,800 = 10,440 above absolute zero ===> 10,400 - 460 ~ 9,800 F.

EXPANSION

In general, as materials gain heat, they expand, and as they lose heat they contract. This is due to the molecules within the object moving further apart as the substance obtains more thermal energy. Proportionally, gases will expand or contract more than liquids during temperature changes of the same amount, and liquids will expand or contract more than solids in the same situation. This expansion is how a thermometer displays the temperature. As the thermometer gains heat, the mercury or alcohol within the bulb will expand. The device is calibrated in such a way that it expands a measured amount, falling to a marker that shows the temperature of the environment when the thermometer reaches equilibrium. This expansion is also the reason that warm air rises over cooler air, since the gases will become less dense as they heat.

Water, however, does not follow this rule. Instead, solid water is actually less dense than liquid water. Ice will contract as it is heated, reaching its maximum density at 4°C, and then begin to expand again. This is why ice floats on top of the liquid in a glass of water, and why ponds and rivers freeze from the top down. The phenomenon of floating ice is an excellent example of how the

What Is Absolute Zero?

Absolute zero is a theoretical measurement of the lowest possible temperature. At this temperature, atomic movement reaches its absolute minimum point, and all lateral movement stops. An object at absolute zero contains no heat energy at all, and cannot transfer any more energy to other objects. While this temperature is only theoretical and cannot actually be reached, scientists have come very close.

In 2003, a group of MIT researchers cooled sodium atoms to just 45 ten-billionths of a degree above absolute zero. At these cold temperatures, atoms start to defy the laws of conventional physics, behaving more like waves than particles and forming new states of matter called condensates. "At such low temperatures, atoms cannot be kept in physical containers, because they would stick to the walls. Furthermore, no known container can be cooled to such temperatures. Therefore, the atoms are surrounded by magnets, which keep the gaseous cloud confined. 'In an ordinary container, particles bounce off the walls. In our container, atoms are repelled by magnetic fields,' explained MIT physics graduate student Aaron Leanhardt in a September 11, 2003, issue of *MIT News*."

For reaching the record-low temperatures, the MIT researchers invented a novel way of confining atoms, which they call a gravito-magnetic trap. As the name indicates, the magnetic fields act together with gravitational forces to keep the atoms trapped.

Researchers study materials at these temperatures because it allows them to take a closer look at atoms while they are moving more slowly, in the same way that it would be easier to describe the features of a car traveling ten miles an hour instead of sixty.

properties of physics in matter affect all other branches of science: The layer of floating ice on top of a body of water protects the life beneath it and allows for a much wider variety of aquatic life at latitudes with cold winter temperatures.

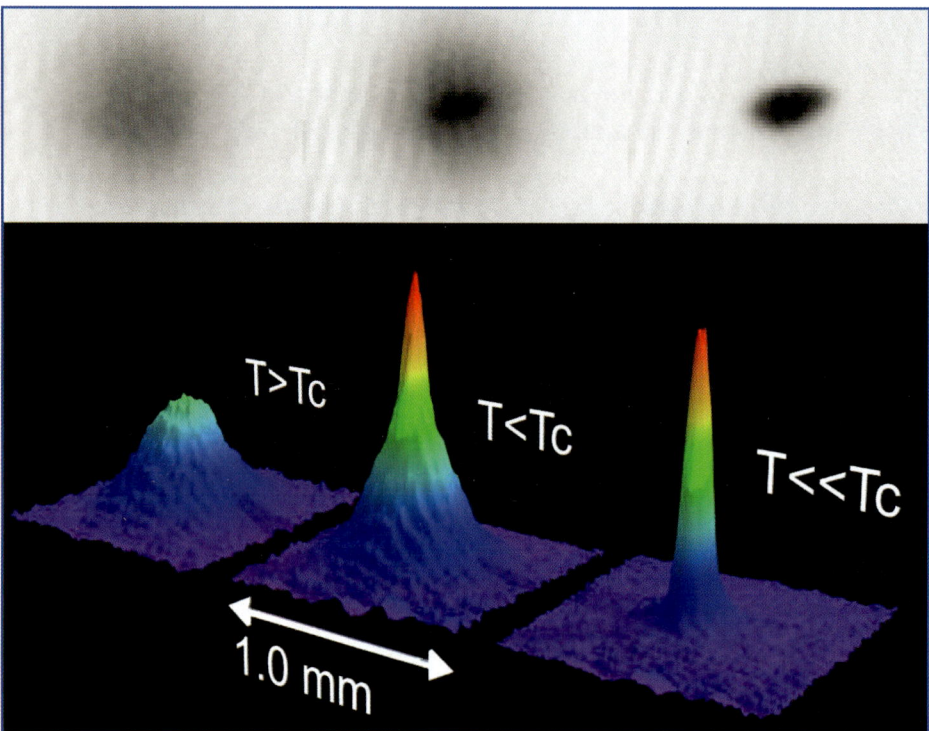

Observation of Bose-Einstein condensation by absorption imaging. Shown is absorption versus two spatial dimensions. The top row shows shadow pictures, which, in the lower row, are rendered in a three-dimensional plot where the blackness of the shadow is represented by height. The "sharp peak" is the Bose-Einstein condensate, characterized by its slow expansion observed after 6 msec time of flight. The left picture shows an expanding cloud cooled to just above the transition point; middle is just after the condensate appeared; right is after further evaporative cooling has left an almost pure condensate. The width of the images is 1.0 mm. The total number of atoms at the phase transition is about 700,000; the temperature at the transition point is 2 microkelvin. (Ketterle Group, MIT)

Wolfgang Ketterle's observation of Bose-Einstein condensation in a gas in 1995 and his first realization of an atom laser in 1997 were recognized with the Nobel Prize in physics in 2001 (together with E. A. Cornell and C. E. Wieman) (Ketterle Group, MIT)

Besides water, other materials will follow the general rule of expansion and contraction according to temperature changes. However, different materials will expand at different rates even at the same temperature. This is the concept behind a household thermostat, which contains two types of metal sandwiched together into a single bimetallic strip, with one type of metal on each side. The different rates of expansion and contraction of the two types of metal cause this strip to bend in one direction as the room cools that closes a circuit to turn on the heat. As the room warms, the strip bends in the other direction, eventually breaking the circuit to turn off the heat.

Bimetallic Strips 1 and 2. The bimetallic strip found in a thermostat has three states—straight, bent up, bent down. The different metals react differently to heat and cold, bending the strip in different directions as the temperature in the room changes. (Department of Physics, California State University, Stanislaus)

SUMMARY

Heat is one portion of the internal energy of matter and relates to the speed at which the molecules within that matter are traveling. The feeling of heat is the transfer of this energy between two objects that are in thermal contact; when those two objects reach the same temperature, they are said to be in thermal equilibrium. Temperature is a measurement of heat. The temperature scales Fahrenheit and Celsius are based on the freezing and boiling points of water, while the Kelvin scale is based on absolute zero. Temperature changes in materials lead to expansion or contraction. Gases expand more than liquids, which expand more than solids. Water is the exception to this rule, as it reaches maximum density at 4°C and is therefore denser as a liquid than a solid. Other materials will expand as they heat, but at different rates depending on the specific material.

2

Temperature and Kinetic Theory

The kinetic theory of matter is responsible for most of our understanding about how matter interacts with other matter in the universe. It is also known as *molecular kinetic theory* or simply *kinetic theory*. In its most basic form, the **kinetic theory of matter** states that all matter is made up of small particles, what are known now to be atoms and molecules, and that these particles are in constant motion. An understanding of the existence of these particles and their movement helps explain such phenomena as *heat transfer*, when heat energy is transferred from one material to another (discussed further in chapter 3), and *Brownian motion*, the visible randomized movement of microscopic and **macroscopic** particles.

The application of molecular kinetic theory to the concept of heat is the basis for such greater concepts as the *ideal gas law*, which is in turn related to the determination of specific heat of gaseous elements and the calculation of internal energy of materials. This chapter explains the basics of the kinetic theory of heat. It also expands on some of the ideas discussed in the previous chapter, of what we perceive as heat being the transfer of molecular

Brownian Motion

Brownian motion, named after the Scottish botanist Robert Brown (1773–1858), is the name given to the presumably random drifting of particles suspended in a fluid (a liquid or a gas) or the mathematical model used to describe such random movements, which is often called a particle theory. In 1827, Brown was studying pollen grains suspended in water under a microscope when he first observed the jittery motion of minute particles ejected from the pollen grains. Although through subsequent experiments he was able to determine that the movement was not related to the fact that the particles were living cells, Brown was never able to determine what was causing the movement. As with many scientific discoveries, contributions by numerous scientists led to Albert Einstein (1879–1955) and Marian Smoluchowski (1872–1917) ultimately articulating the mathematical explanation for Brownian motion in 1905 and 1906, respectively. These discoveries finally confirmed, without a shadow of a doubt, the physical (not just hypothetical) existence of molecules and atoms. The concepts of Brownian motion have been usefully applied to many diverse fields of study, including economics, music, theater, art, biology, and, of course, physics.

motion and the concept of temperature as a measurement of the velocity of this motion. This chapter will discuss the theories behind these concepts as they apply variously to gases, liquids, and solids, and some of the real-world implications of those theories, such as gas pressure and evaporation.

KINETIC ENERGY

The lateral velocity, or forward movement, of a particle, as previously discussed, relates to the heat energy of that particle. The velocity of a particle can also be called the **kinetic energy**, or energy of motion, contained by that particle. As a result of molecular movement equating to both heat and kinetic energy, the temperature of the object can be measured not only by the average velocity of the particles within an object, but also by the average of the kinetic energy of those particles. The kinetic energy of a particle can be measured using the equation $KE = \frac{1}{2} m * v^2$, where KE is the kinetic energy measured in joules, m is the mass measured in kilograms, and v is the velocity measured in meters per second. As a result of the kinetic energy being related to the square of the velocity, doubling the velocity of a particle will result in that particle having four times the kinetic energy.

MOLECULAR MOTION AND HEAT IN IDEAL GASES

Described in its simplest form, the kinetic theory of heat in ideal gases relates the microscopic behavior of molecules in an ideal gas to the macroscopic relationship of that gas to other matter. An **ideal gas** is a theoretical concept that denotes a gas made up of randomly moving point particles. In an ideal gas, these particles do not interact with one another unless they actually physically collide. While ideal gases do not exist in the real world, most gases, such as those commonly found in breathable air, including oxygen and carbon dioxide, act similarly enough to these assumptions that they can be treated as if they were ideal gases, at least for most purposes.

In order to understand how the kinetic activity of molecules relates to heat, it is necessary to understand the kinetic behavior of these molecules. For the purposes of this chapter, when the word *molecule* is used, be aware that either atoms or molecules would behave in the same way, and the word *molecule* is being used alone for the sake of brevity. One aspect of ideal gas behavior includes the assumption that while the number of molecules in a volume of gas is quite large, the space between molecules is also quite large when related to the size of the individual molecules. The result of this large and empty space between molecules means that there are no attracting or repulsing forces acting between them.

Also, any of the collisions between molecules in an ideal gas are **elastic**. An elastic collision means that when two molecules collide, they bounce off of each other traveling in the opposite direction. While the energy of motion can be transferred from one molecule to the other, changing their velocity, during an elastic collision all energy is conserved between the two particles. While truly elastic collisions only exist in theory, just as the whole ideal gas only exists in theory, most gases behave similarly enough that

Temperature, Volume, and Pressure

As has been discussed, gas molecules can be contained inside a vessel; and as the number of gas molecules inside the vessel increases, the number of collisions with the walls of the vessel also increases. The force of these collisions is referred to as the **pressure** of the gas in the vessel. If the vessel in which the gas is contained is capable of expansion, such as in the case of the rubber in a tire or a balloon, adding more molecules will force this expansion and result in an increase in volume. Since increasing the temperature of a gas speeds up the molecules, a change in temperature will also result in a greater number of collisions between the gas and the walls of the vessel. Therefore, increasing the temperature of the gas would then increase either the pressure or the volume of the vessel. Conversely, if you have an object with a constant volume and you decrease the pressure, the temperature will also decrease.

The connection between a change in temperature and a change in pressure can be pictured through the example of a kettle heating on top of a stove. As the water heats and turns to gas molecules in the form of steam, an event that will be discussed further in chapter 4, the water vapor molecules increase in number within the kettle. The pressure builds within the kettle until it is high enough that the water vapor molecules in the form of steam escape through the small opening in the spout, in effect increasing the volume of the container to that of the entire kitchen.

the ideal gas assumptions including elastic collisions hold true for most calculations. The molecules in an ideal gas also move very quickly in random directions, although each molecule moves at a constant speed if no energy is added to the gas. This high rate of speed, and therefore the quite large number of collisions between the gas molecules and the walls of a container, explain how gases under pressure are able to exert force on objects. For example, air can inflate a bicycle tire and make it resistant to impact due to these collisions. As more air molecules are forced into the tire, the collisions of air molecules against the rubber of the tire occur with more and more frequency. Eventually, more collisions occur between the molecules of air inside the tire and the rubber of the tire than occur with the molecules outside the tire, so it feels rigid.

However, since the molecules in a gas are far apart relative to their size, the collisions between individual gas molecules do not occur very often. The kinetic energy of molecules is transferred during collisions, in the same way that energy is transferred from a cue ball to one of the other balls in a game of pool. Since heat is kinetic energy, this means that heat in a gas transfers through molecular collisions, and then since these collisions do not occur very often, heat takes a long time to diffuse through a gas. These collisions are actually an extension from the assumptions used for the ideal gas law, because in that law the particles do not have a definable size but are a single-dimensional point; molecules in reality do have a measurable diameter and do collide with one another.

The calculation of the frequency of these collisions is done using the theory of the mean free path, which is the average of the distances traveled by the gas molecules before they collide with another molecule. This is represented with the *Boltzmann equation*, $L = 1 / (\sqrt{2} * \pi * (n/v) * d_m^2)$ where L represents the mean free path in meters (m) and n/v is the number of molecules per unit volume; there is no unit for n/v as it is a count value, not a measurement; d_m is the diameter of the molecules in the gas, which is also measured in meters (this value changes depending on the element and the atmospheric pressure the gas is under).

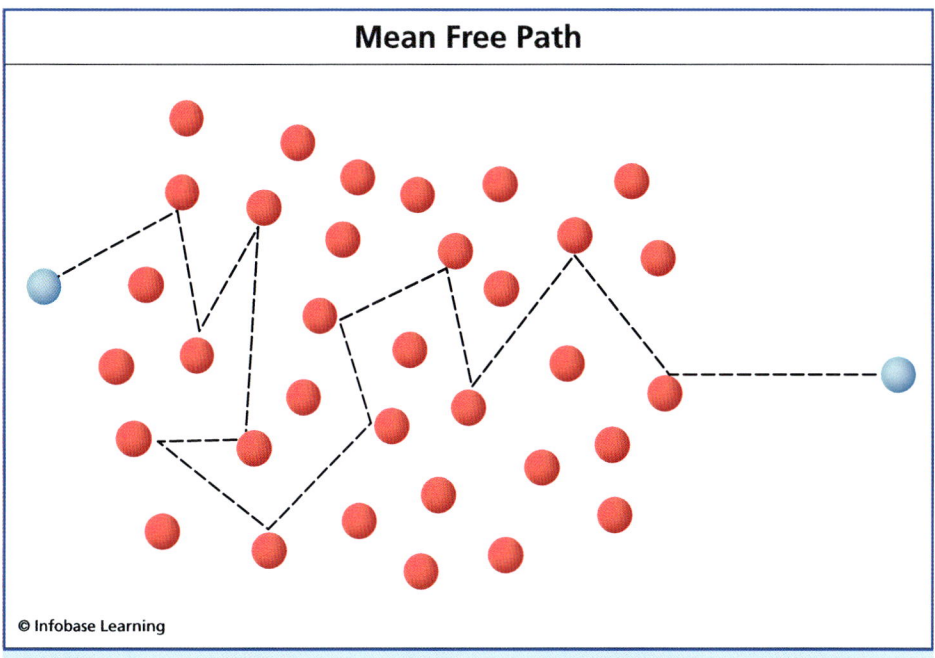

Mean Free Path

All particles moving through space suffer from collisions. In this diagram, the blue dots represent a photon traveling through space and the red dots are matter (protons, neutrons, and electrons) that the photons bounce off of. When they collide, energy is transferred from the photons to the other particles (red dots). The "mean free path" is the distance the photon (blue dot) can travel before colliding with another particle. In higher density matter, the mean free path is shorter. (Source: Department of Physics, University of Oregon)

KINETIC THEORY APPLIED TO SOLIDS AND LIQUIDS

As discussed in chapter 1, the average lateral velocity of the molecules in an object is measured as the temperature of that object. However, this is only exactly true in gases, which have the molecular kinetic properties discussed previously. The properties and molecular behaviors of solids and liquids obviously differ significantly from those of gases, and even more so from ideal gases.

The molecules within solids and liquids are much closer together than those molecules in a gaseous state. In a gas, as discussed, the space between molecules is so large as to allow the molecules to move with almost complete freedom. In materials in a liquid state, the molecules can still move freely within a container, but they must push past one another. Imagine people walking down a crowded city street: They can get where they wish to go, but they will constantly need to push their way through the crowd. The closeness of these molecules means that while the molecules move easily, the attractive forces between them force them to conform

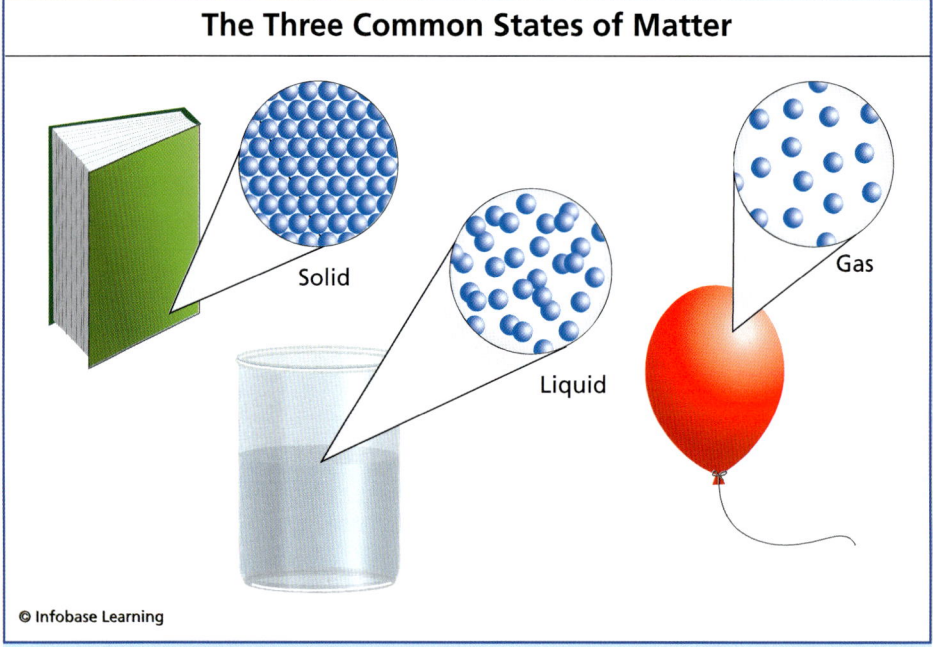

Illustration of molecules in a solid and liquid. While the relative closeness of atoms in a liquid when compared to a gas results in more collisions and faster energy transfer within the material, molecules in some solids are so tightly packed into formation that they cannot move and therefore do not often collide with one another. (Source: Department of Physics, California State University, Stanislaus)

to the shape of a container. In a solid, the molecules are packed even more tightly together, like people standing in a crowded theater pit.

Since the molecules are close together in liquids, they collide with one another more often than do molecules in a gas. This means that heat energy can diffuse through a liquid relatively

History of Kinetic Theory

The kinetic theory of heat came about in the mid-19th century, shortly after the discovery of the laws of conservation of energy and thermodynamics. One major point of resistance to the theory was that it just seemed too simple; could it really be true that gas molecules moving through space have essentially no forces acting on them until they collide with something? Rudolf Clausius (1822–88), a German physicist, was the first to take this idea seriously, publishing a paper on the topic in 1857. Some implications of Clausius's work included the idea that molecules could contain more than one type of atom, which we know to be true in the case of molecules such as oxygen (O_2), and that the total space taken up by the gas molecules must be very small when related to the total volume of a gas. Clausius's paper also described the characteristics of the molecules in the various states of matter, changes of phase, and how all of those things were related to his kinetic theory of heat.

When errors were found in his assumption that the "particles" making up a gas were infinitesimal, Clausius compensated by correcting his theory in a later paper to say that the "sphere of action" for a gas molecule was large enough that collisions would occur with some frequency. He also created the mean free path equation. Luckily for the future applications of that equation, it was immediately taken up by other researchers and applied to kinetic theory their own way; otherwise, it might just have been taken as Clausius grasping at straws to make his kinetic theory of heat work.

quickly. However, the increased rate of collisions due to the close-ness of molecules does not carry over into solid materials. The attractive forces in a solid are so strong that the object must con-form to a specific shape and cannot shift into another formation without considerable added energy. Even though they are very close together, the molecules in solids are set into a rigid arrange-ment that prevents much movement, and this rigidity prevents them from colliding with one another. In many cases in a solid, the crystalline structure holding the molecules together prevents the molecules from moving individually off one another, as if the molecules were characters on a foosball table. The individual char-acters on the table do collide with one another, but only rarely. In a solid material, the rarity of those collisions would prevent the rapid transfer of heat energy. This is why the wooden handle of a pot or cooking utensil will remain cool to the touch even when in contact with a hot liquid.

EVAPORATION

The kinetic theory of heat helps explain the evaporative property of liquids, evaporation being the familiar tendency of liquids to disappear when left out in a dish or to dry out of clothing hanging on a line. Temperature, as we have mentioned, is simply a mea-surement of the average velocity of the particles in an object. If a group of molecules is in liquid form, the average kinetic energy of those molecules is less than is needed to escape the surface tension of that liquid and become a gas. However, this is only an aver-age. By definition, some of the particles will be traveling slower than that average and therefore contain less energy, while some others will be traveling faster than average and therefore contain more energy. Some of these molecules may even be traveling fast enough to evaporate, which means they contain enough energy to shift into the gaseous state of the material and escape the surface of the liquid.

As a result of **evaporation**, the temperature of the liquid in question is reduced. This is because, again, temperature is a mea-surement of average kinetic energy. If the molecules with the highest energy successfully escape from the liquid, the average energy of the group of molecules remaining is reduced, due simply to the mathematical properties of an average. This is why sweating cools your skin; as the drops of perspiration evaporate, the molecules that remain behind in your skin have less energy and the temperature of your body is subsequently reduced. However, the liquid in a bowl sitting out on a table will still fully evaporate until there is none remaining, despite losing heat every time a molecule evaporates. This is because the air in the room remains at room temperature, considered to be at approximately 25°C. The air in the room will transfer heat to the water in the bowl, and bring it back up to this temperature. The faster-moving particles resulting from this heating process will eventually be able to escape

Steam from hot coffee is an everyday example of liquid molecules moving fast enough that they turn into gas, in the form of steam, and escape. (iStockphoto.com)

from the bowl, and the process repeats until all of the liquid has escaped.

SUMMARY

The kinetic theory of matter allows a connection to be made between the behavior of individual molecules and the observable behaviors of visible matter, as well as connecting theoretical equations to real-world experiences. Applying this theory to heat then gives an understanding of the reasoning behind such concepts as the differing rate of heat transfer in various states of matter and the relationship between temperature, volume, and pressure. The assumptions of kinetic theory for gases are based on the concept of an *ideal gas*, which describes a gas made up of noninteracting particles that collide with one another elastically and travel at a constant rate in a random direction. Since the distance between molecules in a gas is large, heat diffuses slowly through it. Heat diffuses most rapidly through liquids due to its closely packed but free-moving molecules; solids are denser but the molecules are generally more constrained and cannot transfer heat energy easily. One implication of the relation of temperature to average kinetic energy is the propensity of liquids to evaporate, since some molecules will travel fast enough to become gas even if the liquid as a whole is too cool.

3

Heat Transfer

One of the major implications of the kinetic theory of matter discussed in the previous chapter is an understanding of the molecular mechanisms of heat transfer. **Heat transfer** is the transfer of thermal energy from one object to another or within a solid object. This can be achieved through the processes of conduction, convection, radiation, and mass transfer. These methods are all related to transfer of heat down a temperature gradient, from an area of high temperature to an area of relatively lower temperature, and can all occur in the same system at the same time. They also all relate to molecular movement and interaction within an object. This chapter will discuss the above-mentioned four major methods of heat transfer, the theories that are behind and inspired by each of these processes, and some examples of their actions and effects in the everyday world.

CONDUCTION

As was discussed more completely in chapter 2, heat energy is actually the kinetic energy that is contained in moving molecules, and temperature is an average of the measurement of this movement. In conduction, the heat energy is transferred through the physical contact of the molecules within each object. As the molecules of the two objects interact, thermal energy is transferred from the molecules of the hotter object to the molecules of the cooler object, which is called **conduction**. These interactions do not need to be actual collisions between two molecules; any type of molecular-level interaction that transfers energy will suffice. Due to this requirement of molecular interaction for conduction of heat, transfer of heat from one object to another through conduction can only occur between objects that are in physical contact with one another. Conduction is also the method by which heat energy travels through a single object or system, as when a solid block of material is heated on one side and the other side of the object becomes warmer. A significant majority of heat transfer in solids is performed through the process of conduction.

Since conduction requires molecular interaction, it makes sense that the less dense an object is and the further apart the molecules in the object are, the less efficient conduction becomes in transferring energy. Conduction as a heat transfer method therefore is most effective in solids, less effective in liquids, and the least effective in gaseous materials, where the molecules are relatively far apart when compared to solids or even liquids. Similarly, as it also relates to the amount of molecular movement and the number of molecular interactions, objects become more efficient conductors of heat as their temperature increases. This temperature increase results in faster moving or more excited molecules that interact more frequently, so the heat energy can be transferred between molecules more easily.

The rate of heat conduction is described through *Fourier's law*, sometimes called the law of heat conduction; however, this law only applies to conduction within solids that are homogenous, or the same all the way through. This law states that the rate of heat transfer as a function of time is proportional to both the surface area of the object that is at a right angle to the direction of the heat flow and to the temperature gradient that exists along this same path. The equation representing this law is

$(\Delta Q/\Delta T) = -k * A * (\Delta T/\Delta x)$, where $(\Delta Q/\Delta T)$ is the rate of heat transfer, measured in watts (W); k is the conductivity of the material, measured in $W * m^{-1} * K^1$; A is the cross-sectional surface area of the object, measured in square meters (m^2); ΔT is the temperature gradient across the material, measured in kelvins (K); and Δx is the distance between the two ends of the distance of the heat gradient being studied, measured in meters (m).

The calculations made possible by Fourier's law allow researchers to determine the proportional ratio for the transfer of heat in a particular material, represented as k in the equation. The

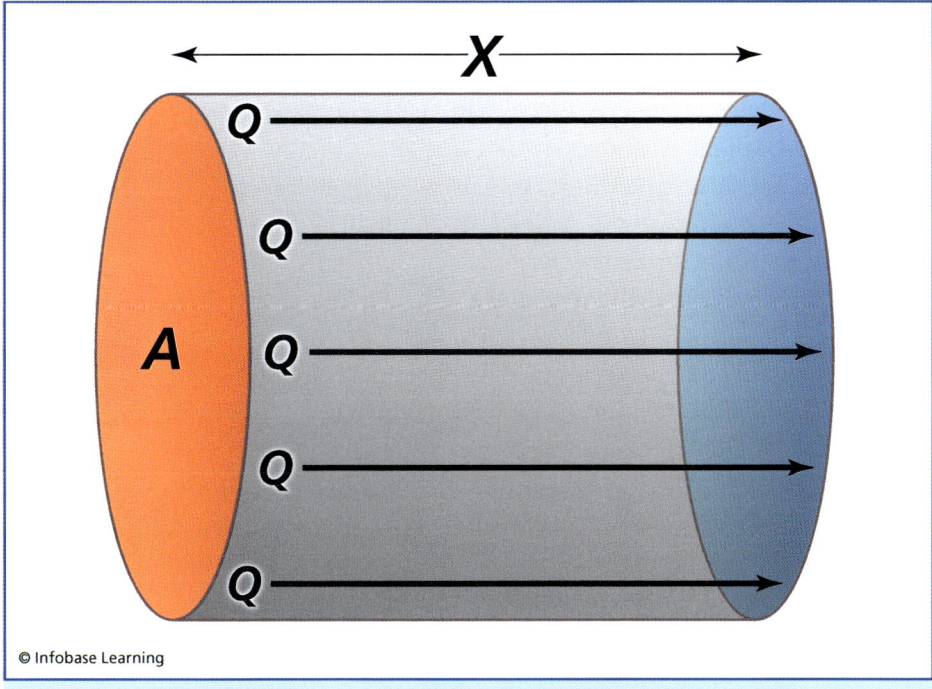

A diagram that shows the flow of heat from hot to cold, to illustrate the law of heat conduction (Fourier's law): Q is the amount of heat transferred; A is the cross-sectional surface area; and X is the temperature difference between the two ends. (Source: Wikimedia)

Jean-Baptiste Joseph Fourier

A French mathematician and physicist, Jean-Baptiste Joseph Fourier (1768–1830) is best known for his investigations into heat transfer and vibrations and for the development, subsequently, of Fourier's law. He is also credited with making the first postulation of the greenhouse effect. In addition to his scientific and mathematical achievements, Fourier accompanied Napoleon Bonaparte (1769–1821) to Egypt in 1798 and became his friend and advisor.

property this proportion represents is also known as the **thermal conductivity** of that material. This is a useful property to know as the thermal conductivity of a material describes how well that material can conduct heat; metals or other materials with a high level of thermal conductivity will conduct heat away from other objects, while materials with a low level of thermal conductivity, such as foam or fabric, make effective insulators. This is why objects made of metal feel cold to the touch. The situation is not that the object is actually colder than other objects in the same room, but that the molecules in the metal are so effective at transferring the heat energy away from your hand that you perceive a temperature difference. As discussed in chapter 2, solid materials are generally less effective at transferring heat than liquids or gases, but the degree of this is going to depend on the formation of the solid. This formation, how the molecules are held together, is what decides the thermal conductivity of a solid.

Conduction is probably the form of heat transfer that we are most familiar and comfortable with in our day-to-day lives since on a macroscopic level it is just energy transfer through physical contact. Touching a hot surface and the resulting burn on your hand is heat transfer through conduction. Holding an ice cube in your hand as it melts is also an example of conduction. In both

cases, heat energy is being transferred from the relatively hot object into the relatively cooler object through molecular interaction
between two objects that are touching each other.

CONVECTION

Heat transfer through convection is dependent on the actual
movement of molecules in a *fluid*, which is any material that de-
forms under a sheer stress, such as liquids and gases. Convection
is actually made up of two processes: mass transfer and heat diffusion. The heat diffusion portion has the same mechanisms as
conduction, with the added property that the molecules can move
freely and collide with one another: random molecular vibrations
result in molecular collisions, and energy is transferred from one
molecule to another during these collisions. The mass transfer

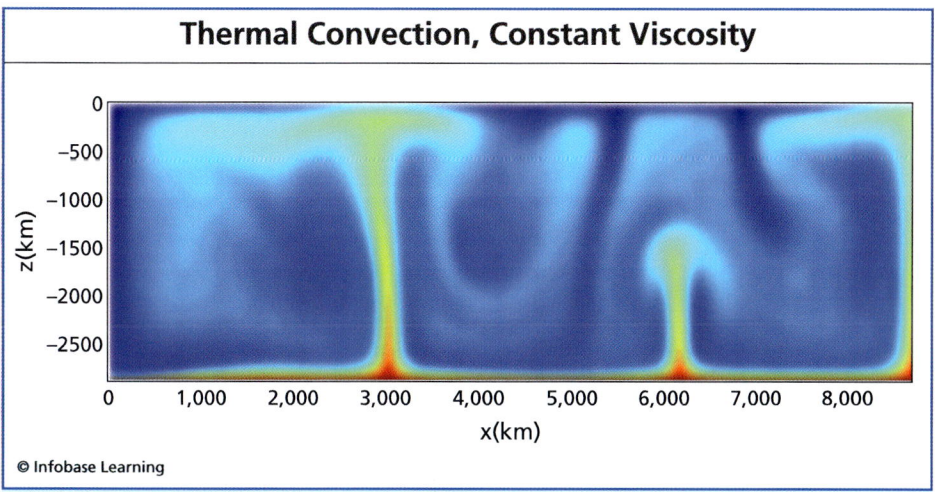

*In this figure, a hot, less-dense lower boundary layer sends plumes of hot material
upwards, and likewise, cold material from the top moves downwards. Colors closer
to red are hot areas and colors closer to blue are cold areas. This figure is based on a
model of convection in the Earth's mantle that shows a calculation for thermal convection.* (Source: Wikimedia)

portion of the process consists of the movement of fluid molecules from one portion of the material to another. Molecules moving in groups carry the heat energy and cause a temperature gradient in the fluid, of which the result is a heat transfer that is called **convection**. This transfer of molecules will result in currents within the fluid, as the hotter, less dense, fluid rises and forces the cooler, denser, fluid downward; these currents ensure that the fluid will

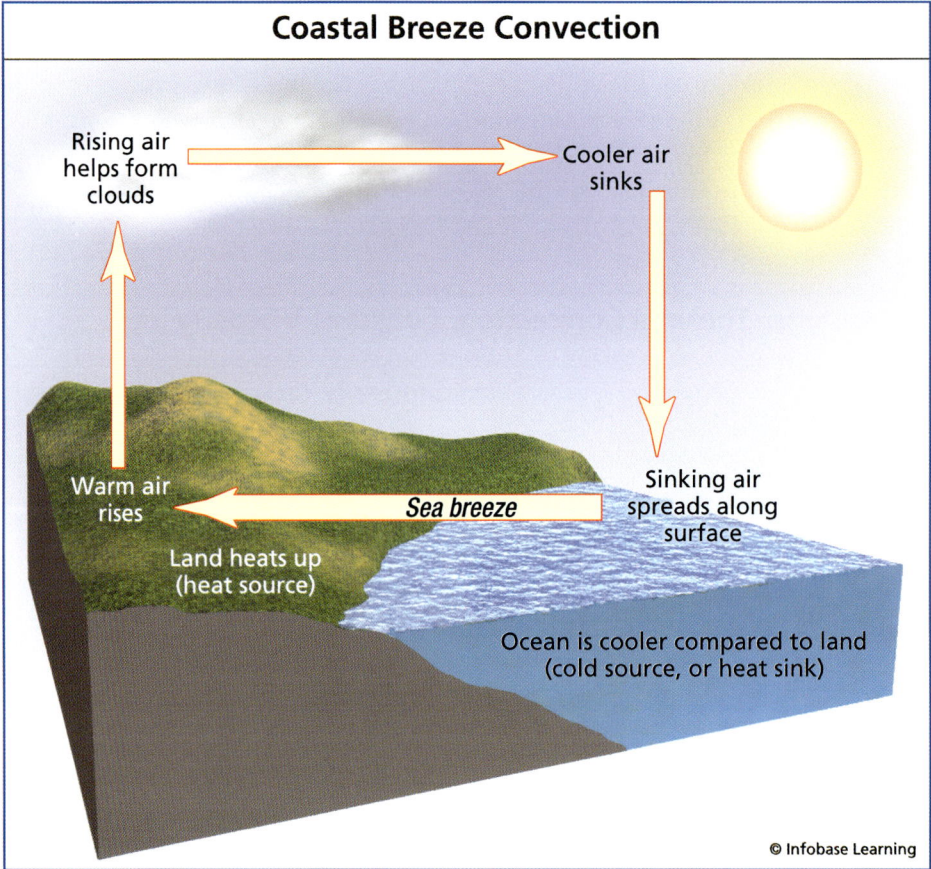

Coastal Breeze Convection

Rising air helps form clouds

Cooler air sinks

Warm air rises

Sea breeze

Sinking air spreads along surface

Land heats up (heat source)

Ocean is cooler compared to land (cold source, or heat sink)

© Infobase Learning

This is a figure of coastal breezes that shows the process of convection and the resulting currents that cause coastal weather systems to operate the way they do. (Source: Brisbane Hot Air Ballooning)

become heated to the same temperature all the way through once the source of heat energy is removed from contact with the fluid.

Convection cannot occur within solids, since the kind of molecular flow required for mass transfer is not possible in molecules that are tightly bound together in solid form. However, the process of convection can transfer heat from a solid material to a fluid, or vice versa, if heat is transferred through a boundary layer that forms in the fluid next to the solid. Heat is then diffused through this boundary layer into the solid through the process of conduction.

Examples of convection generally involve situations where you can see part of a fluid move into another area. For example, a steaming cup of coffee will heat the air above it through the process of convection, as the molecules in the steam move from the liquid to the gas and take the heat energy with them. One result of convection on the boundary layers between solids and fluids is coastal weather systems. Since water is more resistant to temperature changes than land, due to the property of specific heat discussed in chapter 1, it takes longer to heat during the day but also longer to cool back down at night. As a result of this temperature gradient, breezes blow from the high temperature to the low temperature. Therefore, during the day breezes blow from the land to the ocean, while at night the breezes blow from the ocean to the land.

RADIATION

Convection and conduction require physical contact between objects; radiation, however, can heat objects that are far apart with nothing connecting them. Any matter above absolute zero, which as we learned in chapter 1 is all matter in the known universe, contains some thermal energy, which it emits in the form of thermal radiation. This type of heat transfer is known as **radiation**. Because this energy is emitted in the form of *electromagnetic radiation*, radiative heat transfer can occur even through the vacuum of empty space. This means that two objects can be in thermal contact, a condition of two objects discussed in chapter 1, even if they are not touching or interacting with each other in any way other than the radiative transfer of heat energy.

Despite differences in the actual ability of the transfer process to move heat without physical contact, the mechanism of radiative heat transfer works because of the same basic molecular properties as conduction. Thermal radiation is a result of the random motion of molecules in an object. Atomic theory teaches that atoms are made up of three types of subatomic particles: protons, electrons, and neutrons, of which protons and electrons are involved in radiation. When the positively charged protons and the negatively charged electrons in the molecules interact, energy is released in the form of electromagnetic radiation. Some of this energy is in the form of thermal radiation, which falls in the infrared portion of the electromagnetic spectrum; when this type of electromagnetic radiation interacts with another object, heat energy is transferred. The theory behind the levels of radiation given off by an object is called the *Stefan-Boltzmann law*, which states that the amount of radiation emitted is proportional to the temperature of the object. The Stefan-Boltzmann equation is $E = s * T^4$, where E is the amount of energy radiating from the object, measured in watts per square meter; T is the temperature of the object, measured in kelvins; and s is a proportionality constant known as the Boltzmann constant with a value of 5.67×10^{-8} watts m^{-2} K^{-4}.

An object at *thermal equilibrium*, that is, an object that is not changing temperature over time because no thermal energy is being added or lost from the object, is also going to be in **radiative equilibrium**. This means that the amount of radiation energy that is being absorbed by the object is equal to the radiation being output by that object, and so the temperature and the amount of radiation remain constant. For an object that is not in radiative equilibrium, if the amount of radiation energy being put into the system increases and becomes greater than the radiation being put out by the object, the object will become warmer; if the amount of radiation decreases and becomes less than that being put out, the object will become cooler.

Ludwig Boltzmann

Ludwig Boltzmann (1844–1906), for whom the Stefan-Boltzmann law is named, is best known for his development of statistical mechanics, which describes how atomic properties, such as mass and charge, decide the perceptible properties of matter, such as viscosity and diffusion. In addition to the Stefan-Boltzmann law, Boltzmann's name is associated with the Maxwell-Boltzmann distribution law, which describes how the energy of a gas is distributed among its molecules; Boltzmann's equation, which expresses entropy in terms of probability; and Boltzmann's constant, a factor used in his equation.

Ludwig Boltzmann, whose name is associated with many fundamental principles and theories in physics, is also credited with founding the field of statistical mechanics (Karl Franzens Universitat Graz)

The most commonly encountered example of heat transfer through radiation is the thermal energy portion of the light of the Sun. One of the unique features of radiative energy transfer is that the thermal radiation can be focused using mirrors or other reflective surfaces, allowing many of the industrial uses for the energy produced by the Sun that are in use today. For example, solar power can be created with a concentrating generator, which uses mirrors to aim the thermal energy and heat water. Another example of radiation is the heat you feel from sitting near a campfire, as the energy radiates from the molecules in the burning logs and agitates the molecules in your skin. The energy from the campfire must obviously be transferred to you through radiative transfer, since the molecular interactions required in conduction or convection mean that you would have to be sitting in the fire!

MASS TRANSFER

Mass transfer is probably the simplest form of heat transfer. As discussed before in the section on convection, heat energy can be transferred through the physical movement of molecules from one area to another. This can be as mundane as using a kettle of hot water to heat a lukewarm bath or using an ice cube to cool

Black Bodies

Due to the properties of the electrons in the molecules, a material that emits high levels of heat through radiation will also be able to absorb heat the same way. There is a special type of material that is referred to as a **black body**, even though it may not be black in color, which is both a perfect absorber and a perfect emitter. In practical terms, this means that all radiation that hits the object is absorbed, while the object also simultaneously gives off radiation energy. It is referred to as a black body because at ambient temperatures such as those found on Earth, the object does not radiate in the visible spectrum and so appears black. However, the wavelength of the emitted radiation changes with the temperature of the black body object, as stated in the Stefan-Boltzmann law, and at higher temperatures the radiation given off is at a much shorter wavelength, eventually reaching the range of visible light. The Sun is considered to be a black body, emitting radiation across the electromagnetic spectrum while also absorbing all the radiation that strikes its surface.

a hot drink. In the first example of adding hot water to a tub of cooler water, the faster-moving molecules in the hot water collide with the slower-moving molecules of the cool bath water, speeding them up and spreading the energy throughout the whole tub. Once the tub reaches equilibrium, it will be warmer than before, due to the addition of the extra energy and therefore an increase in the average energy and the total kinetic energy of the water. In the second example, the addition of the cold ice cube lowers the average molecular movement of the hot drink. The molecules within the drink collide with the molecules in the ice cube, transferring some of their energy to the ice. This heats up the ice, while simultaneously reducing the energy in the drink. The transference of energy to the ice will also result in melting, or *changing phase*, which is a concept that will be further discussed in chapter 4. When this system reaches equilibrium, the drink will be slightly cooler than before the ice was added, since the average energy of the molecules in the drink and their total kinetic energy have both been decreased.

MPEMBA EFFECT

In reality, of course, none of these heat transfer methods acts alone. Heating and cooling processes are always a combination of some or all of these transfer methods. Sometimes the combined effects of the different types of heat transfer can result in highly unexpected results. One prime example of this is the **Mpemba effect**, the phenomenon of hot water freezing in a shorter time than an identical amount of cold water. If this seems nonsensical to you, you are not alone; the reasons behind the Mpemba effect are still unclear.

The Mpemba effect was studied by historical scientists such as René Descartes (1596–1650) and Aristotle (384–332 B.C.E.), but it was not taken seriously by modern scientists since it seems to contradict what is known about heat under kinetic theory. It was not until 1969, through the work of a high school student in Tanzania named Erasto Mpemba (1950–), that the theory was brought back to the attention of modern physicists. While making ice cream, Mpemba noticed that his bucket of heated milk froze faster than his classmates' buckets of cooled milk. Though his physics teacher told him this was impossible and Mpemba must have been mistaken, he soon discovered that ice cream makers in nearby cities followed the same practice to get frozen ice cream sooner. He convinced a college professor, Dr. Denis Osborne, visiting the school to attempt the experiment. When Osborne's results coincided with his own, a paper was authored and published under both their names, resulting in the name Mpemba effect.

Scientists as yet have been unable to fully explain the mechanism behind the Mpemba effect. It is a combination of a variety of heat transfer methods, and those combinations change under different conditions of water and ambient temperature. Since there is no one mechanism under differing conditions, it is difficult for scientists to pin down exactly what is happening at a molecular level. Most likely, the different mechanisms work together in all cases, with some more important under certain conditions than others. Compounding the difficulty is the fact that the Mpemba effect occurs only under some conditions: water at 210.2°F (99°C) will obviously take longer to freeze than water at 33.8°F (1°C), even despite this effect. The Mpemba effect is most pronounced in a mid-range of temperatures, for example with a cooler sample at 86°F (30°C) and a warmer sample at 158°F (70°C), though disagreements exist in the scientific literature about the conditions under which it will occur.

The first mechanism thought to explain this phenomenon is evaporation. Another scientist, Dr. Kell, published a paper at the same time as the Osbourne-Mpemba paper discussing a similar effect and attributing the result to evaporation of the hot water and subsequent loss of mass leading to the hot water freezing first. While evaporation and the resulting loss of mass is related to the Mpemba effect, and may be the single most important factor, the effect will still occur even in a closed container that does not experience a change in mass. In addition, other experiments have measured the mass loss due to evaporation and found that it was not sufficient to explain the difference in freezing times.

It is also possible that the use of boiled water for the hotter temperature sample may be related to the Mpemba effect. Boiling a sample of water causes it to release dissolved gases, meaning that as it cools it contains less gas than a sample of water that starts at a cooler temperature. Speculations about this property range from claims that the lack of dissolved gases increases the possibility of convection currents and increases the speed of heat transfer through convection to claims that it changes the enthalpy of freezing of water (a property that will be discussed in chapter 4). While experimentation does hold some of these claims to be true, there seems to be no theoretical basis that can explain them.

Another possibility relates to the properties of heat loss through convection. As discussed before, heat transfer through convection results in convection currents, cycles of fluid in the system that carry heat energy. Since the molecules in warmer water are more spread out than those in cooler water, warmer water is less dense than cool; convection currents result in the water being stacked in order of density, with the warmest on top. In a freezer, this hot water on top is exposed more quickly to the cold ambient temperatures and could therefore cause the sample to lose heat energy faster than the cooler sample, which would have fewer convection currents and a less-pronounced stacking effect. Once again, however, this alone does not explain the Mpemba effect.

Other complex explanations have been proposed to explain this effect. For example, it is possible that the material of the containers provides different insulation at different temperatures, or that the hot water could melt frost on the surface of the freezer and provide a better conductor for the loss of heat energy from the container to the walls of the freezer. It is also suggested that the cooler sample might be more prone to **supercooling**, or freezing at a temperature lower than its typical freezing point, than the hotter sample; this would result in the warmer sample freezing first since it would actually be freezing at a higher temperature. Despite any claims made in the scientific literature, however, no single explanation exists for the Mpemba effect that explains the effect in all conditions. Until enough experimentation is done to cover all possible scenarios that result in the Mpemba effect, it will simply remain a mystery of modern science.

SUMMARY

Relating the kinetic theory of matter to the concept of heat gives us an explanation for the behavior of molecules with regard to heat transfer, which is the movement of thermal energy from one object or portion of an object to another. This can be accomplished through the major methods of *conduction, convection, radiation*, and *mass transfer*. All four methods are related to the propensity of molecules to move randomly. Conduction is most common in solids and is the result of interactions between molecules. Convection is most commonly seen in liquids and is actually a result of a combination of the effects of heat diffusion, or energy transferred through molecular collisions, and mass transfer. Heat transfer through radiation is a result of the electromagnetic energy produced from the molecular movement in all matter above absolute zero. Mass transfer is the simple transference of molecules from one place to another, resulting in a change of temperature in the second system. Interesting results occur in the real world when these methods coincide, such as the Mpemba effect, where warmer water freezes faster than cooler water. Even with an understanding of heat transfer methods and kinetic theory, no single explanation yet exists for the Mpemba effect. No heat transfer method alone explains the results of the Mpemba effect.

Change of Phase

As we learned in the previous chapters, adding or removing heat from a system changes one property of that system: the average velocity of the molecules in the system, which is measured as temperature. As heat is added to or removed from the system in the form of energy, the temperature will rise or fall steadily—that is, until it reaches a critical point, known as the **phase transition point**. At this point, the matter begins to shift into a different form, which is referred to as a *phase* or a *state* of matter. The shift itself is referred to as a **phase change** or a phase transition.

This chapter will cover the four major states of matter and their properties, which are necessary background to understanding phase changes between these states. The types of phase changes that occur and the necessary conditions for a phase change are discussed. The concept of enthalpy and its role in the mechanisms of a phase change is also addressed.

STATES OF MATTER

In order to discuss the properties of the changes between states, it is necessary to first understand the properties of the states of matter themselves. There are four main states of matter: solid, liquid, gas, and plasma, and a small number of additional and unusual states. Each state has its own set of properties and behaviors,

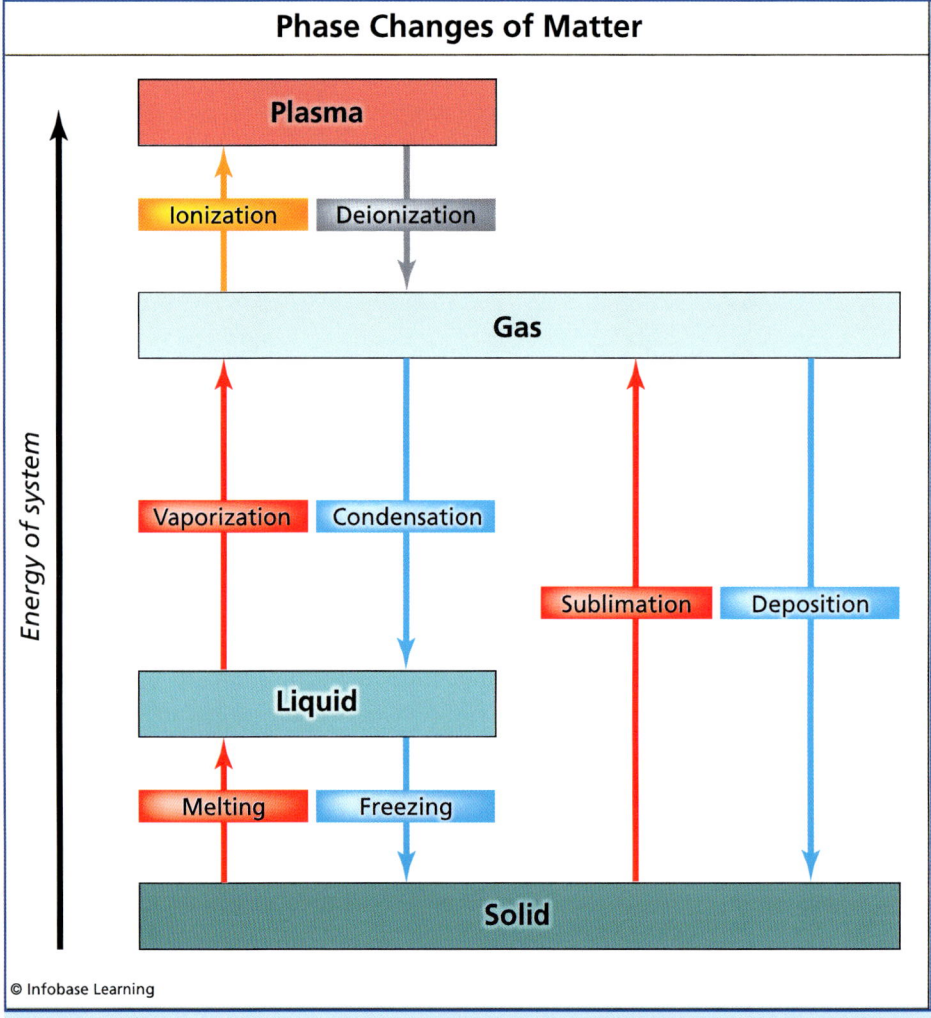

There are four major states of matter, which under the right conditions can shift from one phase to another

which cause a material to behave drastically differently when it changes from one to another.

Solids have a definite volume and a definite shape, and the atoms in the material are fixed in position relative to one another. Solids will not change their shape based on the shape or size of an enclosure. They are the only state of matter than can resist **deformation**, which means they can withstand a pushing force without changing shape. **Liquids** have a definite volume but an indefinite shape. This means that a liquid takes the shape of the container it is poured into, but the total amount of liquid remains the same. The atoms in a liquid move relative to one another but are close together, making them, as was discussed previously, good conductors of heat energy. **Gases** have no definite volume or shape. A gas is simply a loose collection of molecules that have no bonds between them. A gas will spread out to fit both the shape and the volume of a container. If it is not in a container, a gas will expand infinitely out from its point of origin. A good example of this is the Earth's atmosphere, which becomes thinner the further from the Earth's surface it travels, but theoretically never fully dissipates. **Plasma** has similar properties to gases, but has high levels of electrical conductivity. This is due to plasma being made up of molecules that hold so much energy the electrons spin off and become a separate part of the plasma cloud. The molecules that make up the plasma cloud are **ions**, atoms with a positive electric charge, and the free electrons, with a negative charge. Due to the loose electrons and ions in a plasma, it can be controlled using a magnetic field, which is not true of a gas. While plasma is the least familiar state of matter to most people, it is actually the most abundant in the universe. This is due to the conditions that exist in outer space, primarily the low pressure, being favorable to the creation of plasma.

The following table shows the major phase changes for solids, liquids, and gases. An important point to note is that while an object is undergoing a phase change, the added energy will not change the temperature of the object.

Phase Changes for Solids, Liquids, and Gases

DESCRIPTION OF PHASE CHANGE	TERM FOR PHASE CHANGE	HEAT MOVEMENT DURING PHASE CHANGE
Solid to liquid	Melting	Heat goes into the solid as it melts.
Liquid to solid	Freezing	Heat leaves the liquid as it freezes.
Liquid to gas	Vaporization, which includes boiling and evaporation	Heat goes into the liquid as it vaporizes.
Gas to liquid	Condensation	Heat leaves the gas as it condenses.
Solid to gas	Sublimation	Heat goes into the solid as it sublimates.

TYPES OF PHASE TRANSITIONS

The type of phase change that occurs depends on the state of the matter before the temperature change and the state afterward. **Freezing** occurs when a liquid becomes a solid due to heat energy loss, while **melting** occurs when a solid gains heat energy and becomes a liquid. This happens, for example, in an ice cube tray in the freezer: the liquid water becomes solid ice. When you remove the ice cube to put it in a drink, it will slowly melt and return to a liquid state. When a liquid gains even more energy and becomes a

Contrails, or vapor trails, are condensation trails and artificial cirrus clouds made by the exhaust of aircraft engines or wingtip vortices that precipitate a stream of tiny ice crystals in moist, frigid upper air. (Wikimedia)

Water boiling and turning to steam illustrates the process of vaporization. (IntheKitchen withKath.com)

gas, it is said to be undergoing **vaporization**, like when water boils and turns to steam. The reverse of this, going from gas to liquid, is called **condensation**. The fog formed by your breath if you blow on a cold window is the result of the water vapor in your breath becoming a liquid when it hits the cold glass. A gas becoming a plasma is undergoing **ionization**, while plasma returning to a non-electrically charged gas state is undergoing **deionization**.

Not every element or material will pass through every stage, and most elements have a specific combination of temperature and pressure where these types of phase transitions are most likely to follow an unconventional path. **Sublimation** is when a solid changes directly into a gas, such as when a piece of ice appears to steam at room temperature, while **deposition** is when gas becomes a solid. Frost, where the water vapor freezes out of the air and onto vegetation, is an example of deposition. Frost does not require the liquid phase of water in the form of rain in order to occur. In both these cases, the defining characteristic is that the

Condensates and Superconductivity

As the temperature of a material is cooled to nearly absolute zero, the laws affecting the behavior of its particles start to break away from those in classical physics. Under just the right combinations of temperature and pressure, an element such as hydrogen can shift into a new phase that is not listed in the major types of matter. This state of matter is called a **condensate**. Depending on the type of element used, the atoms could change phase into either a Bose-Einstein condensate or a Fermionic condensate. Unlike the phase changes between solids, liquids, and gases, there is no temperature under standard atmospheric pressure that will cause matter to shift into the condensate phase. Instead, matter must be cooled under very high or very low pressures to reach this state.

One of the major reasons why research into condensates is so important and desirable has to do with a special property of condensates: **superconductivity**. A superconductive material is a perfect transmitter of electricity, meaning it has an electrical resistance of zero. All energy put into the system will be transmitted fully; no energy is lost in moving the electrical current from one part of the material to the other. The applications of this property are numerous. For example, superconductive electromagnets are close to the most powerful types of magnets on the planet. Other applications include digital circuits and proton detectors. While superconductivity can be induced in some materials that are not in the condensate phase, the lower temperature condensate superconductors are generally easier to create than the so-called "high-temperature superconductors."

Water freezing on a window, forming crystalline patterns, illustrates the process of condensation (Wikimedia)

material changes into a gas or a solid without passing through an intermediate liquid phase.

ENTHALPY AND HEAT

Enthalpy is a measurement of the total internal energy of a system, which is the energy contained in the molecules making up the object, plus the amount of energy required for that object to push against the surrounding environment and maintain a stable volume at a given pressure. It can also be measured as the heat energy contained in a system under a constant pressure. The added energy needed to maintain the object at a given pressure is necessary

Dry ice turning into gaseous CO_2. The white vapor seen in this photograph is an example of sublimation as the dry ice, which is actually solid carbon dioxide, becomes gaseous CO_2 without passing through a liquid state (Wikimedia)

due to the collisions between the air molecules in the atmosphere and the object in question. The atmosphere of the Earth actually presses down on every object with almost fifteen pounds of pressure per square inch, a force that must be countered with an internal force; this internal force is added to the heat energy of an object to total the enthalpy of the object. Despite their similarities, enthalpy and heat are two distinct concepts. Enthalpy is measured in joules (J), calories (cal), or British Thermal Units (BTU) and falls under the category of potential energy.

The calculation of enthalpy uses the equation $H = U * p * V$, where H is the total enthalpy of the system, measured in joules; U is the total internal energy of the object being observed, also measured in joules; p is the pressure on the borders and the environment surrounding the object, measured in pascals (Pa); and *V* represents the volume of the object, measured in cubic meters. This equation is used because *H*, the total enthalpy, cannot be measured in itself and must be determined by other properties.

Due to this restriction, most experimentation is more interested in determining ΔH (Δ in physics is commonly used as shorthand for "change in," so ΔH is simply a shorter way to write the phrase "change in enthalpy" during a reaction). ΔH is the difference between the sum of the energies of the *reactants* and the sum of the energies of the *product*. Enthalpy can be changed due to the addition of energy, a change in pressure, or a change in temperature. A reaction with a positive ΔH absorbs heat from its surroundings; this is called an **endothermic reaction**. Conversely, a reaction with a negative ΔH releases heat into the surrounding environment; this is called an **exothermic reaction**. Most chemical reactions are reversible; the enthalpy of a reversed reaction is

The Joule

The joule, a unit of energy, is named in honor of James Prescott Joule (1818–89), an English physicist whose work included contributions to the development of the first law of thermodynamics, the law of the conservation of energy. Joule also collaborated with Lord Thomson Kelvin on the formulation of the Joule-Thomson effect, which describes the way the temperature of an expanding gas cools if the gas does not perform external work. Although Joule never received a formal education, he was interested in physics from a young age and later established a laboratory near his family's brewing business, where he conducted experiments.

The joule is also known as a Newton-meter, as it is the amount of work expended when a force of one Newton is applied to an object over a distance of one meter. However, the Newton-meter is usually used as a measurement of torque, or twisting force, while the joule is used for energy.

−1 multiplied by the enthalpy of the initial reaction. So an endothermic reaction when reversed would have a negative ΔH and become an exothermic reaction, and a reversed exothermic reaction would result in an endothermic reaction with a positive ΔH.

If the pressure on the object remains static and no other work is performed than the expansion of the object, ΔH is the same as the amount of heat added to the object. However, ΔH when the pressure is not stable over time is the combined total of heat added to the object and the nonmechanical work performed on it that results in this change of pressure. **Work** is the action of a force that causes a displacement in the system; nonmechanical work is any work that is not caused either by kinetic energy, the energy contained in a moving object, or by potential energy, the energy stored based on the position of the object. Passing an electric current through the object would be an example of nonmechanical work, as would agitating the reaction components through mixing.

ENTHALPY AND PHASE CHANGES

Previously, we discussed that adding or removing heat energy to an object or system will cause a steady change in the temperature of the object. If enough heat is added to or removed from a system, the material will eventually reach a critical point where it will change phase. The amount of energy required to have that change occur for one mole of the substance is referred to as the enthalpy or heat of the type of change. At this point, the energy absorbed from the environment stops causing a change in temperature and instead begins to change the arrangement of the molecules. The temperature will not increase or decrease further until the phase change is complete. Since the addition or removal of more heat does not have a corresponding increase or decrease in the material, this energy is also referred to as **latent heat**.

There is a distinct enthalpy required for each type of phase change and each element. The **enthalpy of fusion** is the amount of energy required to cause one mole of a substance to move from a solid state to a liquid state. The reverse of this, the energy required to move from a liquid state to a solid state, is called the **enthalpy of freezing**. The **enthalpy of vaporization** is the amount of energy needed for a change from liquid to gas, while the energy for a change from gas to liquid is the **enthalpy of condensation**. Finally, the energy required to move from gas to plasma is the **enthalpy of ionization**, and from plasma to gas is the **enthalpy of deionization**. In all cases, the energy removal to move "down" a state is the negative of the energy required to cause the "upward" change of phase.

The enthalpy of nonstandard phase changes, the change directly from a solid to a gas or from a gas to a solid, can be determined by adding together the energy that would have been required for both phases. That is, that reaction uses both the energy for the skipped phase and the energy for the final phase. This amount of energy is called the **enthalpy of sublimation** or the **enthalpy of deposition**.

The addition of enthalpy in the form of heat is the most intuitive process for reaching the phase transition point and inducing a change in phase. However, in addition to changes in heat, reaching the phase transition point can be achieved through changes in pressure. For example, at very low pressures, materials will remain in a gaseous state at temperatures far below their transition point for condensation or freezing. At very high pressures, matter will remain a liquid at temperatures that under normal circumstances would cause the matter to vaporize. Every element also has a **triple point**, a specific temperature and pressure where solid, liquid, and gas exist interchangeably. These triple points are used to define the size of various base units. For example, the triple point of water is 0.01°C, or 273.16 K; this is the point around which the Celsius scale is calibrated. Water is one of the few types of matter that can be seen in all three phases under Earth-normal conditions, such as steam rising from the ice on top of a partially frozen river. However, this state is due to the agitation of the river water more so than a true triple point state, as the official triple point of water is at a lower pressure than one atmosphere.

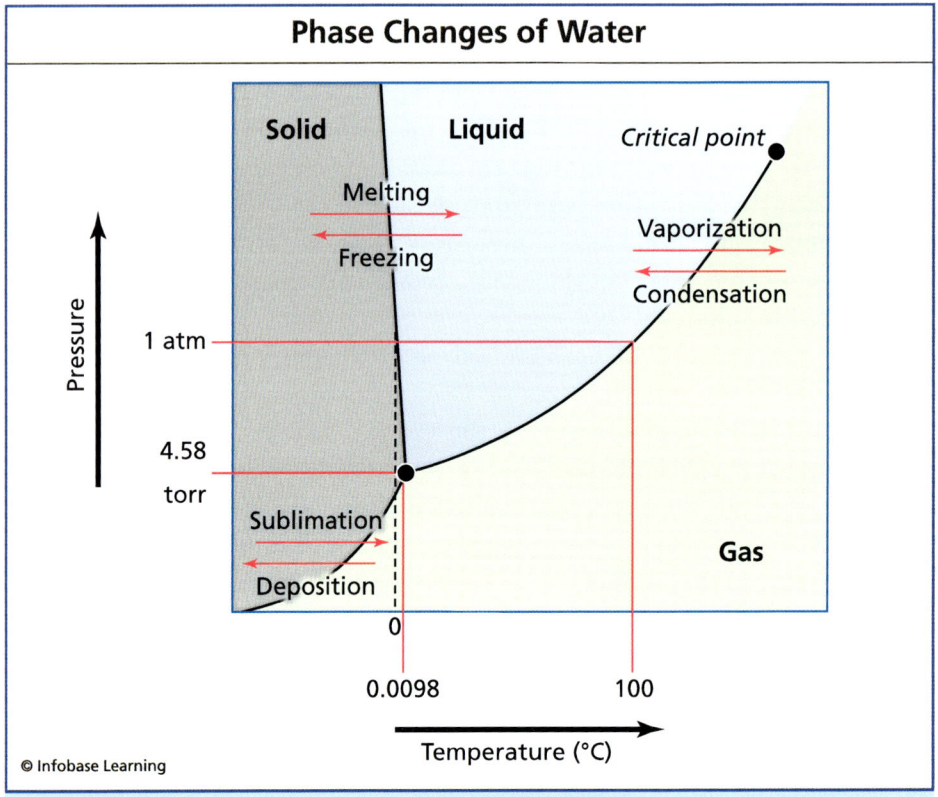

Phase Changes of Water

This graph shows the effect of pressure and temperature on the phase changes of water. It also illustrates the different conditions under which various states of matter can exist. Under normal pressure, H_2O shifts from ice to liquid water at 0 degrees Celsius, but at low pressure will already be in the gaseous phase at 0 degrees Celsius (California Institute of Technology)

SUMMARY

When matter reaches a certain point, known as its *phase transition point*, it undergoes a *phase change*. This means that the matter will shift from one state of matter to another. There are four major states of matter: solid, liquid, gas, and plasma. *Solids* have a definite shape and volume, liquids have indefinite shape but definite volume, and gases and plasma have indefinite shape and indefinite volume. Plasmas are also electrically charged. A shift from liquid to solid is *freezing*, a shift from solid to liquid is *melting*, a shift from liquid to gas is *vaporization*, a shift from gas to liquid is *condensation*, and a shift from gas to plasma is *ionization*. At certain temperatures and pressures, matter may shift directly from a solid to a gas, which is called *sublimation*, or from a gas to a solid, which is called *deposition*.

Enthalpy is a type of potential energy. It is the sum of the energy contained in an object and the energy required to maintain its shape in a pressurized environment, and can be calculated using the equation $H = U * p * V$. However, most researchers are more interested in the change in enthalpy over time, ΔH, which is defined as the difference in energy between the reactants and products in a reaction. *Endothermic reactions* absorb heat and have a positive ΔH, while *exothermic reactions* give off heat and have a negative ΔH. Adding to or removing from the enthalpy in a system generally results in a steady change in temperature until the object reaches its phase transition point. At this point, the energy absorbed or emitted is used instead toward changing the phase of the matter in the object. The amount of energy this requires is referred to as the enthalpy or latent heat of that shift. These include the enthalpy of fusion, enthalpy of freezing, enthalpy of vaporization, and enthalpy of condensation. The enthalpy of sublimation is found by adding together the enthalpy of fusion and the enthalpy of vaporization, while the enthalpy of deposition is found by adding together the enthalpy of freezing and the enthalpy of condensation. This energy requirement can be met either through the addition or removal of heat energy or through changes in pressure.

5

Thermodynamics

Previously, we have looked at the basis of heat and temperature, the mechanisms of heat transfer, and the phase changes of matter under various conditions. The study of these concepts, and especially of heat transfer, their effects on matter, and the work the transfer of heat or the matter itself can do is known as the science of **thermodynamics**. Thermodynamics relates the behavior of heat energy to the effects of forces acting between objects. Unlike kinetic theory, which is concerned with how heat and molecular activity relate, thermodynamics is primarily concerned with observing the large-scale effects of heat transfer. In short, thermodynamics is the study of the physics of energy flow in systems.

CHARACTERISTICS OF THERMODYNAMIC RESEARCH

Only heat transfers that can be broken cleanly into two portions, heat and work, can be studied using thermodynamics; any processes that do not involve one of these two, or involve other sources of energy, cannot be studied using thermodynamics. Any thermodynamic hypothesis or theory must be able to be proven through experimentation or observation, and the protocol for these experiments must allow for a test that would disprove the theory. Like most scientific study, the results of experiments in thermodynamics must be **reproducible**. This means that future researchers must be able to reach the same final product by performing the same experiments, though they may not always draw the same conclusions from this data. In addition, all reactions in thermodynamics must eventually reach a state of thermodynamic equilibrium; the rule of reproducibility requires that the same initial reaction always reaches equilibrium at the same point.

Thermodynamic equilibrium is the point at which the thermal, chemical, mechanical, and radiative properties of a system are stable at a *macroscopic level*, meaning the level of the system as it can be observed without magnification equipment. A system in thermodynamic equilibrium cannot change state or do work, and it will not visibly change over time if it is isolated from other systems or objects. Even systems in a state of thermodynamic equilibrium will exhibit minute fluctuations in their energy levels at a microscopic level, but these changes are simply too small to have a macroscopic effect on the properties of the system. Additionally, in a system at thermodynamic equilibrium, all molecular actions are balanced with an opposite action, a property known as detailed balance. If the initial state and the limits of a nonequilibrium system are known to a high degree of accuracy, it is even possible to predict what the properties of the system will be at equilibrium.

Thermodynamic research is primarily concerned with the study of thermodynamic processes. **Thermodynamic processes** are those reactions that result in a phase change within the subject of study, or reactions that cause the object to exchange energy and matter with the surrounding environment. A special type of thermodynamic process that undergoes several changes in temperature, pressure, and/or state of matter, but that eventually

Thermodynamics and the Flow of Energy

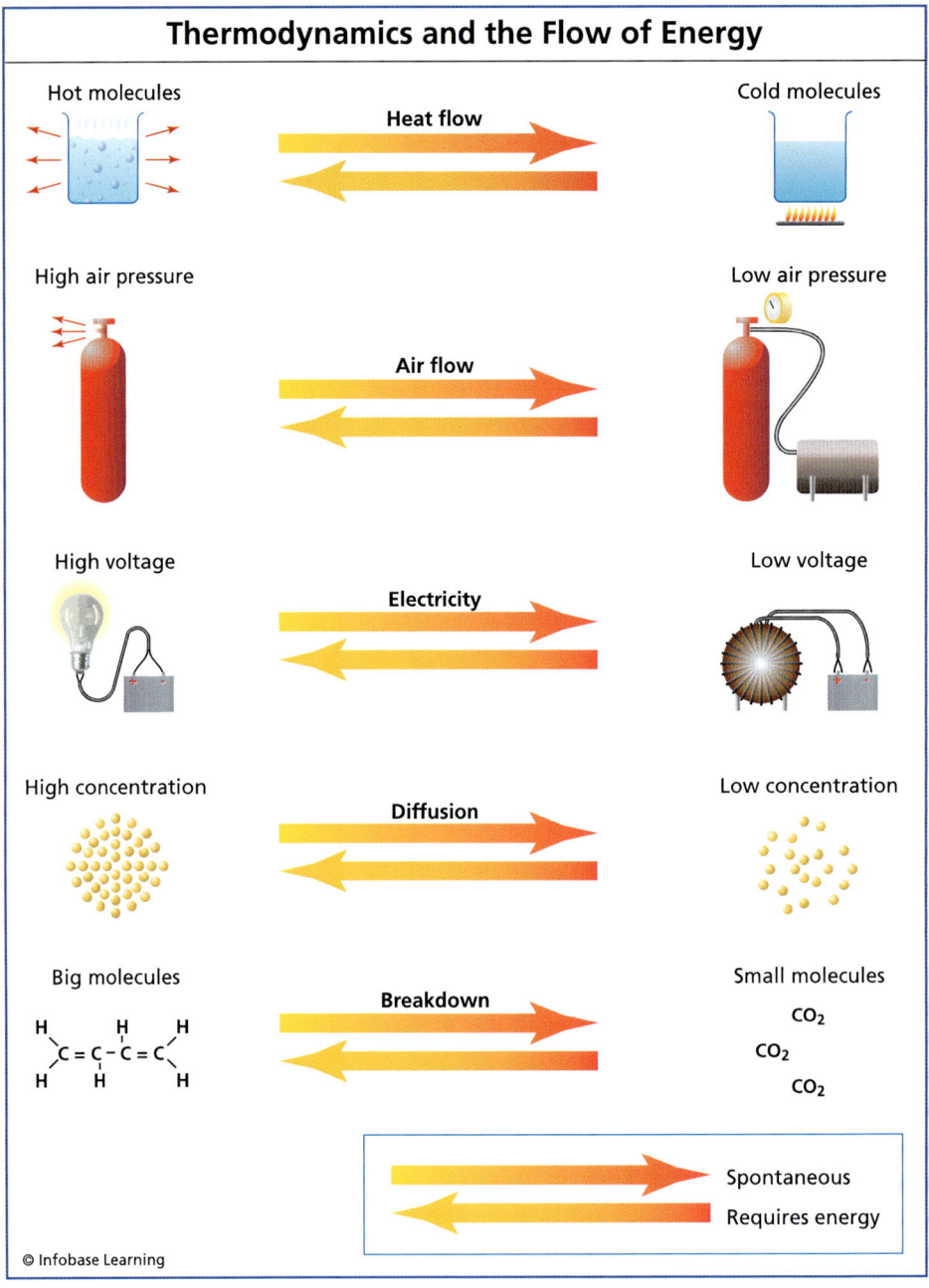

Hot molecules

Heat flow

Cold molecules

High air pressure

Air flow

Low air pressure

High voltage

Electricity

Low voltage

High concentration

Diffusion

Low concentration

Big molecules

Breakdown

Small molecules

CO_2

CO_2

CO_2

Spontaneous

Requires energy

returns to its original state is called a **thermodynamic cycle**. The rules controlling the behavior and properties of thermodynamic processes, systems, and cycles are known as the laws of thermodynamics, and will be discussed in more detail in the following chapter.

THERMODYNAMICS AND WORK

Earlier, work was defined as a force acting on an object that can cause a displacement of that object. **Thermodynamic work**, however, is defined in relation to the energy being transferred in or out of a system, instead of the movement of the object upon which the work is being done. The action must result in a change in the environment surrounding the system to be considered thermodynamic work. The work being performed can cause a change in the temperature and pressure of the system or its environment, the chemical composition of the system, the magnetic or electrical fields produced by that system, or the gravitational effect of the system on its environment.

A fundamental concept in the study of thermodynamics is the difference between *work* and *heat*. Both heat and work use the joule (J) as a possible unit of measurement, but they are very different in their effects on the system on which they act. Heat is a property of the energy in the molecules within a substance, and it can be directly measured through changes in temperature. The

(Opposite page) *Energy and matter change from high energy and highly structured states to lower energy and disordered states. Heat flows from hot to cold places, resulting in a uniform temperature; air flows from high to low air pressure, resulting in a uniform pressure; electricity flows from high to low voltage, resulting in a uniform voltage; molecules diffuse from high to low concentration, equalizing their concentration; and big molecules with much stored energy break down into small molecules with less stored energy.*

transfer of heat is a transfer of molecular energy at the microscopic level. Thermodynamic work cannot be measured directly;
instead, the work is measurable only through its effect on environ-
mental factors. Work, therefore, is not a tangible property in itself
but only a way of thinking about the changes it can perform. Work
also differs from heat in the fact that thermodynamic work can be
observed macroscopically, while of course the transfer of heat energy at the molecular level cannot be observed this way; only the
final effect of temperature change can be measured to evidence
the heat energy transfer. So while changes in heat can be *measured*
directly, and changes in work must be measured indirectly, the
changes in heat are not directly *observable*.

As mentioned earlier, thermodynamics is primarily focused on
the results of thermodynamic processes. These processes are the
methods through which heat energy in the system is converted
into the work performed by the system. There are four types of
thermodynamic processes: adiabatic, isobaric, isometric, and isothermal. Each process may be called by other names as well, as
noted in the following discussion.

In an **adiabatic**, or isocaloric, **process**, no heat energy enters
or leaves the system. The work being performed by the objects
within the system on the system's environment is exactly equal
to the change in the internal energy of the system. This state can
be achieved either through the perfect insulation of the system's
environment, or if the reaction rate is so rapid that there is no
time for heat energy to be transferred between systems. Changes
in temperature in an adiabatic process are due to a change in the
pressure of the system, since as the pressure increases, the speed
and collision rate of the molecules will also increase. A special
subset of adiabatic processes is those that can be reversed—
these are known as isentropic processes; in an isentropic process,
the entropy of the system can be assumed to remain constant
throughout the functions of the process.

An **isobaric** process has an addition of heat that results in both
a change in internal energy in the system and work being performed by the system, but without a change in pressure. With the

pressure held constant, the addition of heat energy results in an expansion of the objects in the system and a corresponding change in volume. In an isobaric process, the change in the amount of enthalpy in the system is equal to the heat energy added to the system. The amount of work performed is the amount of heat energy entering the system minus the change in internal energy.

An **isometric process**, also called an isovolumetric or isochoric process, occurs when heat is added to a closed environment that will not allow the system to expand. This lack of expansion means that the system affected by the isometric process will remain at the same volume throughout the effects of the process. This is the reason for yet another name for this process, a constant-volume process. Isometric processes cannot perform work because all the energy entering the system is used to increase the internal energy; with no expansion possible, the heat energy can only raise the temperature and pressure of the system.

In an **isothermal process** the temperature of the system remains constant. This means that the internal energy of the system will also remain constant, and all of the heat energy can be used to do work. An isothermal process is usually achieved through the use of a heat sink, which absorbs any excess heat energy produced by the system. This differs from an adiabatic process, since in an adiabatic process heat energy can increase the temperature of the system but is not transferred into or out of the system, while an isothermal process may exchange energy with its surroundings but in such a way that the temperature of the system does not change. In an ideal gas, however, adiabatic and isothermal processes both have a net change in energy of zero, since the only contributor to the internal energy of an ideal gas is its temperature. An ideal gas at constant temperature cannot therefore experience a change in internal energy. Another factor to remember about isothermal processes is that the matter within the system can still change its state, since changes in pressure can also have that effect even without a change in temperature.

In addition to the work performed by thermodynamic processes, thermodynamic cycles can also be used to do work. In a

power cycle, heat energy is converted into mechanical energy and can be used to do work. Thermodynamic power cycles form the basis for the operation of heat engines, which are responsible for supplying most of the world's electric power. These cycles create power by converting heat energy, such as the energy produced when a gasoline fuel is burned, into mechanical energy that can be used to drive engines. Power cycles are behind the functioning of automobile engines and electrical plants, which need to put out energy. Gasoline and diesel engines for automobiles, steam turbines used in power plants, and gas turbines used to power airplanes, ships, trains, generators, and even tanks rely on the energy created by thermodynamic power cycles. The reverse of a power cycle, where the temperature is changed between two different states with the energy from a mechanical input, is called a **heat pump cycle**. In other words, instead of using heat energy to do work, heat pump cycles use work to create heat energy. Heat pump cycles are used for refrigerator motors and home heating systems, which are concerned with controlling the temperature of the environment.

The devices that make use of power cycles in order to create mechanical work are known as **heat engines.** Remember, the cycle is the process that draws out the energy, while the engine is the physical system that uses the energy. Heat engines are a popular method for doing mechanical work, since conversion from other types of energy into heat is generally a simple and easy process. For example, the chemical energy contained in the bonds of a hydrocarbon fuel can easily be converted to heat through the process of combustion. Heat engines work by using an energy source, such as a flame, the Sun, or the Earth's heat, to heat up a fuel, the *working source*, to a high temperature. The working source then does work in the engine by giving off heat energy to the *heat sink*, until the working source reaches a low temperature again. Heat pump cycles, which as discussed use the reverse of this process to change the temperature of their surroundings, are also considered to be a type of heat engine.

The amount of energy a heat engine can put off and the efficiency of converting that energy to work are both a function of the difference of temperature between the heated working source and the heat sink. This determination is what is known as Carnot's theorem, and the reasons behind it will be discussed more thoroughly in the next chapter. Most gasoline automobile engines, for example, operate at an efficiency of about 25 percent, while a coal-fired power plant achieves about 45 percent efficiency. The lowest temperature point of a heat engine is limited by the temperature of its environment, and so on Earth the efficiency of heat engines in that direction is limited. Research into the use of heat engines and increasing their available energy focuses primarily on increasing the starting temperature of the working source.

Different types of heat engines rely on different properties of the working source, as well as different engine configurations and different methods for turning heat energy into work. Phase change engines rely on the work that can be performed by the expansion or contraction of the working source as it changes state. Steam engines are an example of a phase change engine; the work performed by a steam engine is done through the expansion of water into steam as it is heated, usually by coal, wood, or gas flame. Gasoline engines also rely on this principle, since it is the expansion of the gasoline into gas in the pistons that produces the work done by such an engine. Some engines instead rely on the expansion and contraction of the working source in a single phase, in a gas-only or liquid-only engine. Still other heat engines rely on the properties and effects of electricity, magnetism, or evaporation in the working source and in the device.

The amount of work that can be performed by a particular system is dependent on the initial state of the system, the final state of the system, and the process through which the work is generated causing the system to take on its final state. The amount of heat energy transferred out of or into the system of interest is also related to the initial state, the final state, and the process that reaches the final state. However, experimentation shows that the final state of a system does not always differ based on the process used to achieve that final state. This is due to the additional variable of internal energy, since any change in internal energy is independent of process and dependent only on the energy levels of the initial and final states of the system. Heat energy and the work performed cannot then be saved by the system, since they are dependent on the process used to create the energy and the work; in contrast, internal energy does not require a specific process and can be stored by the system for release at a later time.

APPLICATIONS OF THERMODYNAMICS

All applications of the science of thermodynamics are based on the ability of thermodynamics to measure and predict the amount of work that can be done in or by a system based on the input of a given heat flow. Most of the research and study in the field of applied thermodynamics is involved with the processes that convert fuels, especially chemical fuel, into heat energy. These calculations are a way to quantify the energy flow between two objects in a system and to determine the efficiency and likelihood of those occurrences.

The quantified data provided by a thermodynamic analysis can then be used to design models of even very complex systems. For example, the branch of science known as atmospheric thermodynamics studies the changes in state and the work performed by water vapor in the Earth's atmosphere. The behaviors of atmospheric water vapor in turn affect such events as cloud formation and wind direction. Models based on equations from atmospheric thermodynamics can predict changes in local weather systems and the formation of tropical storms and hurricanes, which is the basis for modern meteorology. Researchers can also use these models and the study of thermodynamics to show the effects of global climate change on the behaviors of atmospheric water vapor; warmer air can hold more water vapor than cooler air, for example, so a warming of the air results in an increase in humidity and a decrease in precipitation.

Historically, the study of thermodynamics focused primarily on improving the design and widening the possible uses of the steam-powered engine. In 1824, a researcher named **Sadi Carnot** published a paper that connected the concepts of heat and energy to the power of the steam engine. This report was the first pub-lished work to make these deductions; due to this work, Carnot is now considered to be the "father of thermodynamics."

Classical thermodynamics studies thermodynamic processes and equilibrium states through the observation of measurable, macroscopic changes in the properties of systems. Many of the theories based on classical thermodynamics have been shown to

Sadi Carnot

Sadi Carnot's (1796–1832) contribution to physics resulted from the pub-lication of a scientific paper in 1824 entitled *Reflections on the Motive Power of Fire*. Although the significance of Carnot's ideas was not duly recognized until 1849, long after his death from cholera in 1832, his name was given to the theories that evolved from this paper—the Carnot cycle and Carnot's theorem. Carnot, a Frenchman, was motivated by nationalist pride to outdo the British in the development of the steam en-gine. Carnot deduced that the efficacy of the engine depended only on the temperature at the heat source and at the heat sink, not on the tem-perature of the working substance. The first deduction formed the basis of the Carnot cycle; the second deduction became known as Carnot's theorem. This theorem has the same effect in a system as the second law of thermodynamics, though it actually predates the formalization of the laws of thermodynamics.

Later work allowed scientists and engineers to use the science of ther-modynamics to improve the efficiency and power output of the steam engine. Carnot's work laid the foundation for classical thermodynam-ics, which is the description of the states (especially equilibrium states) and processes of thermodynamical systems. Classical thermodynamics describes these states using macroscopic, empirical properties directly measurable in the laboratory. It is used as a method of illustrating the exchange of energy, work, heat, and matter, and is based on the laws of thermodynamics.

be correct even though scientists of that time period did not understand the molecular and atomic activity responsible for the macroscopic changes in the properties of a system. It is called *classical* because it is concerned with macroscopic empirical parameters that can be measured in the laboratory, which was the first level of understanding about thermodynamics and in fact all of chemistry and physics, and the only level of study known in the 19th century. Later, a microscopic interpretation of these concepts was described by the development of statistical thermodynamics. Statistical thermodynamics uses theories of probability to calculate the expected behavior of the particles in a system.

Another application for thermodynamics is the analysis of chemical reactions. If all of the important aspects of a reaction within a thermodynamic system, such as pressure, temperature, volume, and energy, are graphically illustrated, it is then possible to determine, as mentioned before, the equilibrium point of the relevant reactions. Most importantly, illustrating the reaction this way can show researchers whether a specific chemical reaction will occur spontaneously under certain conditions, and under what conditions that reaction will require energy to be added to the system in order for it to occur. The study of chemistry through thermodynamics is also interested in the relationship between chemical reactions and phase changes in matter. Studying the heat flow in a chemical reaction can help scientists understand which reactants are limiting factors and help determine the efficiency of a catalyst for a given reaction.

Modern propulsion engines are only in existence due to the study of thermodynamics. In a jet engine, also known as a power turbine engine, a compressor pumps air into the engine compartment and through a burner, which heats the air. The heated air then passes over the rotors, or blades, of the turbine. The work performed by the hot air on the turbine rotor is what causes the rotors of the turbine to spin. This spin provides the lift needed for the airplane to take off from the runway and remain in the air after that.

Jet propulsion engines were developed as a result of the study of thermodynamics (Wikimedia)

SUMMARY

Thermodynamics is a branch of science that studies the transfer of heat and the work that can be performed by this heat transfer, and focuses on the relationship between the forces acting on objects within a system and the behavior of heat energy. In essence, thermodynamics is the study of energy flow. Research into thermodynamics must follow the same rules as other scientific experimentation, including being testable, reproducible, and disprovable. One feature of reactions that can be analyzed with thermodynamics is that those reactions must eventually come to an equilibrium point, while reproducibility requires the same reactants to come to the same products at the same equilibrium point.

Thermodynamic processes are reactions that result in a change in state or a transfer in heat energy, or both. *Thermodynamic cycles* are a subset of thermodynamic processes that will pass through several stages and eventually return to its original state. These processes are the methods through which heat energy is converted to mechanical work, and can perform *thermodynamic work*, or transfer energy into and out of a system. There are four types of thermodynamic processes: adiabatic, isobaric, isometric, and isothermal. Adiabatic processes have no energy entering or leaving the system, and a special type of adiabatic processes called *isoentropic* processes occur with no change in entropy in the system. Isobaric processes occur under constant pressure, isometric processes occur with constant volume, and isothermal processes occur with constant temperature. In a *power cycle*, heat energy is converted into mechanical work, while in a *heat pump cycle*, mechanical work is used to change the temperature of the surroundings. Power cycles are at the basis of heat engines, which provide much of the world's mechanical work today. Heat engines are seen in vehicles, power plants, refrigeration units, and in many other modern applications. Without them, the modern world as we know it would not function.

Applied thermodynamics focuses on the ability of thermodynamic science to measure and predict heat flow, energy transfer, and work, as well as the methods through which fuel can be converted into mechanical energy. Historically, thermodynamics was designed for and applied to the efficiency of the steam engine. Today, the science of thermodynamics is used to predict the weather and the effects of global climate change, in the analysis of chemical reactions, and in the design and manufacture of propulsion engines.

6

The Laws of Thermodynamics

The laws of thermodynamics are the foundation for the study of thermodynamics. The properties of heat and energy in matter, as well as the transfer of and work done by heat, are defined using the laws of thermodynamics. Any system that involves the conversion of heat energy to some other form of energy or uses heat energy to perform work must obey these four laws. However, unlike most scientific facts, the laws of thermodynamics are so basic that they cannot be proven through derivation from other equations or scientific theories. Instead, the accuracy of these laws has been proven through repeated observation and experimentation, and the laws have thus become the foundation upon which many theories in physics are developed.

Since they were determined prior to the formulation of atomic theory and the kinetics of heat, the laws of thermodynamics do not discuss the mechanisms behind the effects of the laws. They deal strictly with the observable, macroscopic events that result from the universe following these laws, though later scientific research has expanded upon the original statement of the laws using the application of molecular and kinetic theories. This chapter will discuss the realm of influence, the theoretical basis, and the implications for each of the four laws of thermodynamics.

THE ZEROTH LAW OF THERMODYNAMICS

The zeroth law deals with the concept of thermal equilibrium. As we recall, thermodynamic equilibrium and thermal equilibrium are not the same. *Thermal equilibrium* means that while heat energy does shift between two connected objects, there is no net heat flow and the temperature of the objects does not change over time. *Thermodynamic equilibrium*, on the other hand, requires that the object be in equilibrium for all properties.

Essentially, the zeroth law is a physics application of the transitive property of mathematics, the idea that if $A = C$ and $B = C$, A and B must also be equal. The zeroth law states that if system A and system B are both in thermal equilibrium with system C, then systems A and B are in thermal equilibrium with each other as well. In other words, any two systems that are in a state of thermal equilibrium with a third system are both in thermal equilibrium with each other. From a practical standpoint, this means that system A and system B are both at the same temperature, and this temperature is the same as that of system C.

The zeroth law is so named because it is the most fundamental of the four laws but was not officially composed until well after the first, second, and third laws had been accepted by the scientific community as a standard. In fact, it is because the zeroth law is so basic that it was not among the original theoretical framework; up until the 1930s, physicists saw no need to formalize or spell out the reasoning used in the zeroth law. Once it was determined that this law was necessary, it did not make sense to continue with the original numbering scheme and call it the fourth law since the other three laws require the existence of the zeroth law.

Thermometers function due to the implications of the zeroth law. For example, imagine that you are trying to find out the temperature of a beaker of water. System A is the beaker of water, and

Thermodynamic Equilibrium (Zeroth Law)

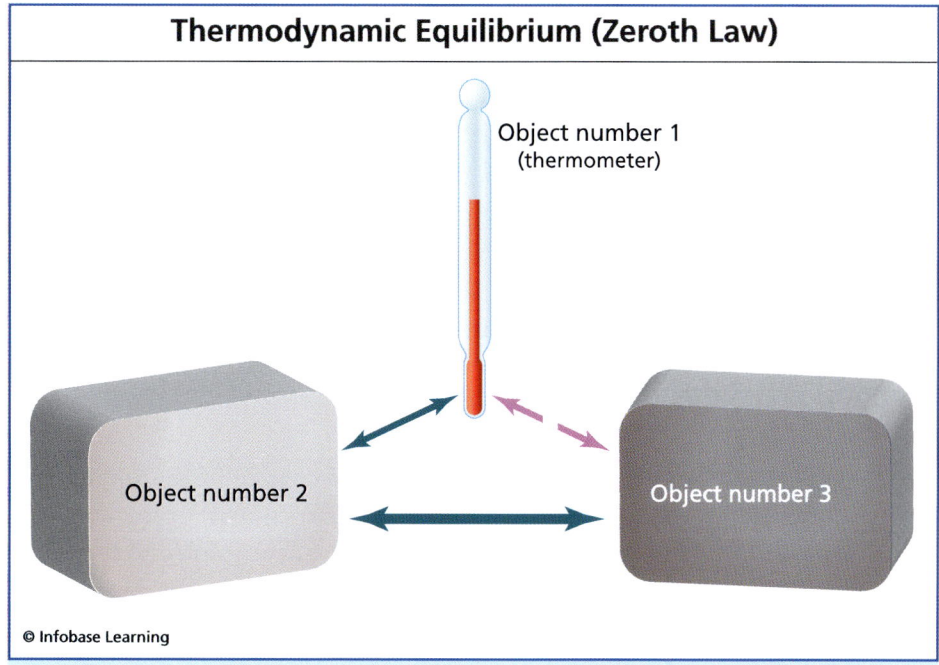

Object number 1
(thermometer)

Object number 2

Object number 3

© Infobase Learning

Temperatures of an inaccessible system can be determined through the use of the zeroth law. The zeroth law of thermodynamics introduces the concept of thermodynamic equilibrium, in which two objects have the same temperature. If two objects that are initially at different temperatures come into physical contact, they eventually achieve thermal equilibrium. During the process of reaching thermal equilibrium, heat is transferred between the objects. The amount of heat -transferred delta Q is proportional to the temperature difference delta T between the objects and the heat capacity c of the object. (Glenn Research Center NASA)

system C is the thermometer. System B in this case is the material that had been used to calibrate the thermometer. Since the calibration with system B shows how the thermometer changes with temperature changes, the level of mercury in the thermometer once it reaches equilibrium with the beaker of water tells us the temperature of the water.

THE FIRST LAW OF THERMODYNAMICS

The first law of thermodynamics is the basis for the principle of conservation of energy. This principle is then related by the first law to the behavior of heat flow and the functioning of thermodynamic processes. Since the first law delineates the parameters of conservation of energy, most branches of science outside of physics state the first law as the principle of conservation of energy. That is, by non-physicists the first law is often said to be the statement that energy can change form and shift between systems, but it cannot be created nor destroyed. The implication of the first law of thermodynamics, when stated this way, is that the amount of energy in the universe is a constant.

The first law states specifically that any change in the internal energy of a system that causes the system to shift from one equilibrium state to another is the same as the heat energy added, minus the work performed by the system. This is usually represented by the equation $\Delta U = \Delta Q - \Delta W$, where ΔU is the change of internal energy, ΔQ is the heat added to the system, and ΔW is the work performed by the system, which causes a drop in internal energy. If ΔW is positive, this indicates that work is being performed *on* the system, thereby increasing the internal energy. This convention exists due to the roots of thermodynamics being in the design of steam engines; steam engines take *in* heat and put *out* work.

This law holds true no matter what type of work is performed by or on the system. Any type of work will cause the system to reach the same final equilibrium temperature by gaining or losing the same amount of heat energy, as long as the total amount of work performed remains the same. In this way, the first law applies to chemical reactions such as those occurring within biological organisms as well as mechanical systems such as engines. A chemical reaction cannot produce more work than the energy contained by the reactants.

The heat energy and work connection of the first law is behind the design of heat engines, such as the heat pump cycle mentioned previously. The heat pump uses mechanical energy to change the temperature of its environment. According to the first law of thermodynamics, the heat pump can use only the mechanical work that is present in the system to perform this temperature change. This is because the total change of energy cannot be more than the sum of the pumped heat energy and the mechanical work that powers the pump.

Another implication of the first law of thermodynamics is that in an isolated system, one with no heat flow or matter transfer in or out, the internal energy will remain constant. If no work is being performed on or by the system and the system is thermally isolated, the sum of the work and the change in heat is zero. Since the work performed plus change in heat energy is zero, and the principle of conservation of energy tells us that energy cannot simply appear within the system, the change of internal energy must also be zero.

THE SECOND LAW OF THERMODYNAMICS

The second law of thermodynamics deals with the direction of heat flow and entropy. In the same way that the possible amount of work that can be performed by a system is constrained by the first law, the second law limits the efficiency of heat engines. The second law is stated several ways, depending on the application. However, the initial statement, called the Clausius statement due to its inventor, the German physicist Rudolf Clausius, in 1850, states, "No process is possible whose sole result is the transfer of heat from a body of lower temperature to a body of higher temperature." In other words, heat energy cannot move from a cooler object to a warmer object unless work is applied to the system. Therefore, any natural process that involves heat transfer can have only one direction, from the warmer object to the cooler object, and without additional work, these processes are irreversible.

Derived from this initial statement, the second law can also be stated as "the entropy of the universe tends to a maximum,"

Rudolf Clausius, the German physicist and mathematician, was the first to articulate the concept of entropy. (Wikimedia)

meaning that disorder in a thermodynamic system will always increase until the system reaches an equilibrium point. **Entropy** can be defined in a number of ways: as the amount of disorder in a system; the amount of multiplicity, which is the number of possible microscopic configurations; and the amount of energy in the system that cannot be converted into work. In a closed system, meaning that no extra energy is added and no work is performed on the system, this gain in entropy cannot be reversed. Systems will always become more disordered, with more possible configurations of matter, and the proportion of energy that is not available to do so will always increase. The combination of the first and second laws stated this way shows why **perpetual motion** is impossible. The first law tells us that no energy that was not originally present in the system can be created from nothing, and the second law tells us that energy will always be lost from a system in the form of entropy. Without added energy, all machinery will eventually run down from the energy loss involved simply in running itself; a perpetual motion machine cannot run itself indefinitely, and it certainly cannot output work in addition to this.

While it may seem that biological organisms defy the second

Perpetual motion, although impossible, has long fascinated inventors and scientists. This cover of the October 1920 issue of Popular Science *magazine, painted by American illustrator Normal Rockwell, depicts an inventor working on a perpetual motion machine.* (Wikimedia and Popular Science, p.d.)

Ecosystem Energy and Matter Flow

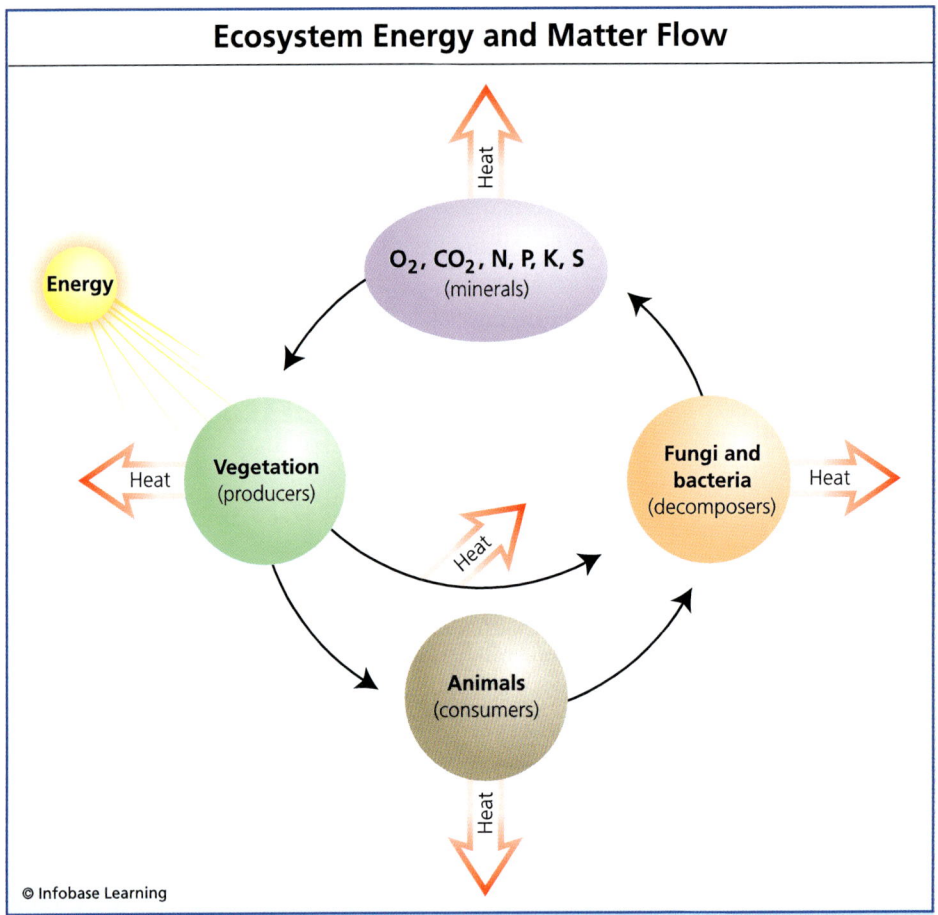

© Infobase Learning

All plant life, animal life, and microorganisms contribute to the continual recycling of the Earth's elements. According to the second law of thermodynamics, energy as heat is lost with each step in the cycle. Organisms that perform photosynthesis capture energy from the Sun and so replenish the energy-matter cycle.

law of thermodynamics, this is not actually the case. Multicellular biological organisms do form order out of disorder, going from a single cell to a bundle of cells and finally to a complete organism. However, this is possible because energy is being added to the system in the form of nutrients for animals or bacteria and sunlight

The Second Law and the Heat Death of the Universe

One of the implications of the second law is that the universe is constantly losing usable energy, since this energy is being used to perform work and is being converted from useful energy into entropy. No process within a closed system, which includes the universe, can reverse a gain in entropy. At the point of the big bang, the very beginning of the universe, the total entropy was zero. This means that the universe was in the highest possible state of order. Since the big bang, entropy has been gradually increasing with the expansion of the universe. According to the heat death theory, the universe will reach a point where matter is distributed evenly across the entire available space, and entropy will reach a maximum point. Once this point is reached, the entire universe will be at thermal equilibrium; without a temperature gradient to cause a heat flow, no work can be performed, which means in effect that the universe can no longer produce light or motion, and certainly cannot sustain any kind of life. This theory about the possible fate of the universe does not, by its name, imply that any particular absolute temperature has been reached. Heat death simply means that temperature differences may no longer be used to perform work. Physics describes this state as maximum entropy.

There are two possible end-state scenarios that involve the "heat death" of the universe. In the first scenario, the expanding universe eventually reaches a critical size and then re-collapses. During this collapse, both the entropy and the temperature of the universe increase to an extremely high level. The final equilibrium state of the universe in this scenario is then densely packed, but still cannot perform work since all the matter is at the same high temperature. In the second scenario, the universe continues expanding indefinitely. As the matter in the universe becomes more and more spread out, entropy increases but the temperature drops. The end state of this expanding universe takes longer to reach than the collapsing universe, and will end with temperatures very close to absolute zero. According to this theory, matter will be spread across the entire universe at an even gradient, unable to interact or perform work.

for photosynthetic organisms. The energy from these nutrients is added to the system of interest, the biological organism, so the second law is still upheld in this scenario.

The heat death of the universe is just one theory about the possible fate of the universe. It revolves around the concept of entropy and posits that the universe will eventually reach maximum entropy, a condition under which life can no longer be supported. (NASA)

THE THIRD LAW OF THERMODYNAMICS

Like the second law, the third law of thermodynamics deals with entropy, in this case entropy as a function of temperature. In its most technical form, the third law states that as the temperature of a system approaches absolute zero, the entropy of a perfect crystal within that system also approaches zero. As discussed earlier, absolute zero is the lowest possible temperature, the point at which molecular movement is at its lowest possible level. A **perfect crystal** is an arrangement of molecules in patterns that repeat exactly the same way throughout the crystal; there may be no deviation from this pattern. In this arrangement, the molecules in the crystal are perfectly balanced, with all forces required to keep them in place having an opposite energy in the molecule. The perfect crystal is at a state of zero entropy, the highest possible level of order.

Due to the nature of heat and temperature, the entropy of a system at absolute zero is determined only by the structure of the molecules within the system, since at absolute zero there is no molecular motion to cause an increase of entropy. The entropy of the system is at its lowest possible point when the temperature of that system is at absolute zero. One implication of this law is that the only situation where it is possible for a system to have an entropy of zero is a perfect crystalline structure at absolute zero.

The implication of this law is that as the entropy of a system approaches zero, the temperature of that system also approaches absolute zero. This is because the molecular motion of the atoms in a system is what determines both the temperature and the entropy of that system. As the molecular motion slows, the structure loses less energy in the form of molecular motion, and therefore has a decrease in entropy. This loss of molecular motion also means that the system is losing heat energy and the temperature of the system shows a decrease. A decrease in entropy cannot be separated from a decrease in temperature.

However, absolute zero and perfect crystals are theoretical concepts. No system can actually be reduced to the point of absolute zero, nor can any system be arranged in such a way as to reach

The concept of a perfect crystal is significant in illustrating the third law of thermodynamics (Sacred Healing Art)

a point of true zero entropy. This is because the nature of atomic structure places a limit on the physical arrangement of molecules within an object. A perfect crystal is mathematically possible, but since the crystal must have no flaws at any level of magnification in order to be considered perfect, some defect is certain to exist in any macroscopic crystal. Because the basis of the third law is in these theoretical concepts, the implications and applications of the law are focused on the effects as a system approaches these points, not on the behavior of the system at absolute zero. In any case, a perfect crystal with zero entropy at absolute zero would not be a very interesting object to observe, as it would perform no work or have any motion of any kind.

SUMMARY

The four laws of thermodynamics provide the foundation for the modern theories of energy, heat flow, and entropy, though they do not in and of themselves provide a reason for the behaviors that follow these laws. The zeroth law of thermodynamics was created last but numbered as the zeroth law as it is the basis for the other three laws; it states that any two systems that are in thermal equilibrium with a third system must also be in thermal equilibrium with each other. The calibration of thermometers works as a function of this law.

The first law of thermodynamics states that a change in the internal energy of any system must equal the heat added to the system minus the work performed by the system; if work is done on the system to create heat, the signs of the equation are reversed and the change in internal energy is the negative of the value of heat removed from the system plus the work performed on the system. An extrapolation of this law is the principle of conservation of energy, which states that energy cannot be created or destroyed in a closed system.

The second law of thermodynamics holds that the direction of heat flow must be from a warmer object to a cooler one, from which is derived the necessary increase of entropy in a system and the requirement of energy to reverse it. The combination of this and the first law shows the impossibility of perpetual motion. The third law of thermodynamics states that the entropy of a perfect crystal at absolute zero will also be zero, though the states of absolute zero, zero entropy, and perfect crystalline structure cannot be reached. The study of the third law is instead on the behavior of systems as they approach absolute zero.

7

Conclusion

The behavior of heat energy is at the very center of the physics of matter. The studies of heat transfer, measurement, and related work are all vital to modern mechanical devices, such as cars, refrigerators, home heating and cooling systems, and electrical generating plants. This text focuses on the basics of these concepts, moving from the defining characteristics of heat to the theoretical basis of its study, and finally to the science of thermodynamics.

All matter in the universe is made up of smaller particles known as atoms, and groups of atoms are referred to as molecules. Increasing the temperature and heat energy of an object is one way to increase the internal energy, or the energy contained within an object. This energy is measured using the units of calories or joules, both of which are units defined by the amount and type of work they can perform. While heat energy will always travel in the direction of a temperature gradient, from the object with higher temperature and higher average molecular velocity to the object with the lower temperature, heat energy can be transferred against the gradient of internal energy. A larger but cooler object will have a higher total internal energy simply because it contains more molecules. Heat energy can be transferred between any two objects that are in thermal contact. Objects that are in thermal equilibrium, however, will pass energy equally between the two without any net heat flow. Thermal equilibrium means that the two objects hold the same amount of thermal energy and are at the same temperature. The specific heat of a substance is a measurement of its resistance to temperature change when heat energy is added.

Heat is the energy of moving molecules, while temperature is the measurement of the average velocity of this molecular movement. Temperature is commonly measured using the scales of Fahrenheit, Celsius, and Kelvin. Higher temperatures indicate a higher average velocity for the molecules making up an object; slower speeds are found in solids, slightly higher speeds in liquids, and the highest speeds in gaseous materials. The Fahrenheit and Celsius scales are calibrated relative to the freezing and boiling point of water; the Kelvin scale is calibrated relative to the point of absolute zero, the temperature at which lateral molecular motion stops and all other molecular motion is at its lowest possible point. Due to the increased molecular motion of an object when its temperature is increased, most matter will expand when heated, making solids the densest form of matter and gases the least dense. Water is an exception to this rule, expanding when it freezes. Water has its highest density at 4°C, when it is in liquid form but only slightly above freezing. Other materials will expand when heated but at different rates depending on the type of material, which is the mechanism behind a room thermostat.

The kinetic theory of matter is the scientific theory that tells us that all matter is made up of atoms and molecules that are in constant motion. The energy of this motion is known as kinetic energy. Temperature, then, can be measured either through the average velocity of an object's molecules or as an average of the kinetic energy of those molecules. This theory was developed using a model known as an ideal gas, a theoretical concept that describes a gas made up of noninteracting point particles that only interact through completely elastic collisions. The space between the molecules in an ideal gas is quite large, even though there is an extremely high number of molecules in the gas. The force of the collisions of those molecules with one another and with the walls of a container is known as the pressure of that gas. Increasing the temperature of a gas will increase either its volume or its pressure; conversely, increasing the pressure of a gas by adding more molecules will increase either the volume or the temperature. The mean free path of the molecules in that gas allows for an estimation of the frequency of molecular collisions, telling us how quickly energy can be transferred within that gas.

Despite kinetic theory being designed for the characteristics of ideal gases, it can also be applied to the behavior of solids and liquids. The molecules in a solid or a liquid are much more tightly packed than in a gas. In a liquid, molecules are close together but can move freely, meaning that heat energy can easily be transferred through a liquid. In a solid the molecules are even closer together; however, in many solids, the structure of the molecules prevents them from moving freely and therefore prevents them from easily exchanging energy. This makes liquids excellent conductors of heat energy while many solids make excellent insulators against it. Since liquids are made up of constantly moving particles, and temperature is measured as an average of the molecular movement for an object, some molecules may actually be traveling much faster or slower than the average. The fastest molecules may even contain enough energy to escape the liquid into a gaseous state, which is known as evaporation.

Heat transfer is the movement of heat energy from one object to another. There are four major methods of heat transfer: conduction, convection, radiation, and mass transfer. Heat transfer through conduction requires the physical transfer of energy through molecular collisions. Convection is a combination of heat transfer through diffusion, which has the same mechanism as conduction, and the mass transfer of gaseous molecules. Radiation involves the transfer of heat energy through the electromagnetic waves that are given off as a result of molecular motion, while mass transfer is heat transfer by moving molecules from one system to another. Objects at radiative equilibrium will maintain the same temperature because they are absorbing exactly as much radiation as they are emitting; black bodies are objects that are perfect absorbers and perfect emitters of radiation, and appear black to the spectrum visible to the human eye. All of these heat transfer methods work together on real-world systems with often complex and intriguing results, such as the Mpemba effect.

When a material is heated to a specific temperature at a given pressure, known as its phase transition point, that material will undergo a phase change. It could shift from solid to liquid, liquid to gas, gas to plasma, or the reverse of any of those processes. Under certain other conditions, matter can shift directly from solid to gas or from gas to solid, or can exist as a solid, liquid, and gas simultaneously, which is known as the triple point of that material. The energy that is added to an object at its phase transition point will not increase the temperature of the object but be used to cause the structure of the object to change phase, and is known as the latent heat of that element. The enthalpy of an object is another way to measure its energy, and is the sum of the internal energy and the energy required for that object to maintain its shape under a known atmospheric pressure. The amount of energy required for a change in phase is known as the enthalpy of that shift, for example the enthalpy of fusion or the enthalpy of vaporization. Reactions that result in the system taking in energy from its environment and increasing their enthalpy are known as endothermic reactions, while reactions that release energy to the environment and decrease the enthalpy of the system are known as exothermic reactions.

The science of thermodynamics is the study of heat transfer and the work that can be performed by those transfers. Thermodynamic work is defined as the ability of a force to cause a temperature change in its environment, as opposed to mechanical work, which is the ability of a force to cause a displacement of an object. Research performed in thermodynamics must produce results that can be reproduced by others performing the same experiments, and any reaction that is being studied using thermodynamics must eventually reach a point of thermodynamic equilibrium, where it is in equilibrium with regard to its chemical makeup, mechanical properties, and radiative effects. However, this equilibrium will only hold at the macroscopic level, or the level of the system as it can be observed without magnification equipment; even systems at thermodynamic equilibrium will undergo microscopic fluctuations. A system at thermodynamic equilibrium does not change over time and cannot perform work. The state of a system when it reaches equilibrium can be predicted by its initial conditions, if those initial conditions are known to a high degree of accuracy.

The study of thermodynamics is mostly concerned with thermodynamic processes, reactions that cause a change in temperature and/or state of the system, and thermodynamic cycles, multistep reactions that will return to their initial states. Thermodynamic processes cause a change in the conditions of their environment, which is a representation of the difference between heat and work. Heat is a molecular property of matter and can be measured directly, while work is measurable only by its effects on the environment of the system performing it. The effects of thermodynamic work can be observed macroscopically, while the transfer of heat energy occurs at the molecular level and cannot be made visible.

The thermodynamic processes at the center of thermodynamic research are the methods through which heat energy is transformed into work. There are four types of thermodynamic processes: adiabatic, isobaric, isometric, and isothermal. Adiabatic processes occur in isolation—no energy is added to or released by the system when the process occurs. The work performed by the process is exactly equal to the change in internal energy of the system. Isobaric processes involve a change in internal energy and work, but occur without a change in pressure, while isometric processes similarly have a change in internal energy and work but the volume of the system is unchanged. Isothermal processes occur without a change in temperature; the difference between an adiabatic process and an isothermal process is that the total energy of a system does not change due to an adiabatic process, while an isothermal process can gain or lose energy in any way that does not result in a change of temperature. The two types of cyclic thermodynamic processes are power cycles, which take in heat energy and use it to perform mechanical work, and heat pump cycles, which use mechanical work to change the temperature of the environment. These cycles are the basis for the mechanical devices known as heat engines. The applications of thermodynamics are many. The initial discoverers of thermodynamics were primarily concerned with steam engines, devices that took in heat and put out work. The science of thermodynamics can also be used to model complex systems, such as climate and weather, or to analyze chemical reactions.

The science of thermodynamics is primarily controlled by the effects of the laws of thermodynamics. These laws focus only on the observable effects, not on the molecular mechanisms of heat and work, though in most cases kinetic theory has been used to explain the "why" of thermodynamics. There are four laws, numbered zero through three. The zeroth law is implied by the other three and was not explicitly written until after the initial three laws, and is an application of the transitive property. The zeroth law states that if system A and system B are both in equilibrium with system C, they are also in equilibrium with each other. All three systems will be at the same temperature. The first law deals with equilibrium states of a system, and says that the change in internal energy when a system changes from one equilibrium state to another is the same as the heat added to the system minus the work performed by that system. The principle of conservation of energy is a result of this law. The second law of thermodynamics states that heat flow cannot travel from a cooler object to a warmer one, and any process that involves heat flow is irreversible unless work is performed on the system. One implication of the second law is the concept of increasing entropy in the universe. According to this concept, the disorder, or entropy, of the universe is constantly increasing and can only be reversed through the addition of work. The third law of thermodynamics states that as a perfect crystal approaches absolute zero, the entropy of that crystal also approaches zero. Because absolute zero is the point at which molecular motion stops, entropy of a system at absolute zero is determined only by the structural arrangement of the molecules in that system. Therefore, the converse of the third law is that as entropy approaches zero, so does the temperature of that system, since the motion of the molecules determines both entropy and temperature.

Each concept within the study of heat is built upon another. A full understanding of the laws of thermodynamics as they are understood today cannot exist without an understanding of the kinetic theory of heat and temperature, and how those relate to heat transfer. Thermodynamic work also connects back to heat transfer methods and phase transitions, since one or both of those changes must occur for thermodynamic work to have taken place. The basics explained within this text should help to facilitate the study of physics, chemistry, and biology, as all science eventually relates back to molecular structure and the transfer of heat.

SI UNITS AND CONVERSIONS

UNIT	QUANTITY	SYMBOL	CONVERSION
Base Units			
meter	length	m	1 m = 3.2808 feet
kilogram	mass	kg	1 kg = 2.205 pounds
second	time	s	
ampere	electric current	A	
kelvin	thermodynamic temperature	K	1 K = 1°C = 1.8°F
candela	luminous intensity		
mole	amount of substance	d mol	
Supplementary Units			
radian	plane angle	rad	π / 2 rad = 90°
steradian	solid angle	sr	
Derived Units			
coulomb	quantity of electricity	C	
cubic meter	volume	m^3	1 m^3 = 1.308 $yards^3$
farad	capacitance	F	
henry	inductance	H	
hertz	frequency	Hz	
joule	energy	J	1 J = 0.2389 calories
kilogram per cubic meter	density	kg m^{-3}	1 kg m^{-3} = 0.0624 lb. ft^{-3}
lumen	luminous flux	lm	
lux	illuminance	lx	
meter per second	speed	m s^{-1}	1 m s^{-1} = 3.281 ft s^{-1}

meter per second squared	acceleration	m s^{-2}	
mole per cubic meter	concentration	mol m^{-3}	
newton	force	N	1 N = 7.218 lb. force
ohm	electric resistance	Ω	
pascal	pressure	Pa	$1 \text{ Pa} = \dfrac{0.145 \text{ lb}}{\text{in}^{-2}}$
radian per second	angular velocity	rad s^{-1}	
radian per second squared	angular acceleration	rad s^{-2}	
square meter	area	m^2	1 m^2 = 1.196 yards^2
tesla	magnetic flux density	T	
volt	electromotive force	V	
watt	power	W	1W = 3.412 Btu h^{-1}
weber	magnetic flux	Wb	

PREFIXES USED WITH SI UNITS		
PREFIX	**SYMBOL**	**VALUE**
atto	a	× 10^{-18}
femto	f	× 10^{-15}
pico	p	× 10^{-12}
nano	n	× 10^{-9}
micro	μ	× 10^{-6}
milli	m	× 10^{-3}
centi	c	× 10^{-2}
deci	d	× 10^{-1}
deca	da	× 10
hecto	h	× 10^{2}
kilo	k	× 10^{3}
mega	M	× 10^{6}
giga	G	× 10^{9}
tera	T	× 10^{12}

PHYSICS PRINCIPLES

The following serves as a reference to provide additional information on important topics and further understanding of the material covered in this book. It includes brief explanations of the laws, theories, and major concepts covered in this book and lists chapters in which the topic is first introduced, although other chapters may also include the topic. Not all chapters are listed because they may not introduce a major theory or concept.

ABSOLUTE ZERO (CHAPTER 1)

Absolute zero, a hypothetical temperature where all movement stops and entropy reaches its minimum, is significant in the study of heat and thermodynamics. It is measured by the Kelvin scale, created by Lord Thomson Kelvin, which is designed so that 0 degrees K is equal to absolute zero. Absolute zero occurs when there is no heat energy remaining to be extracted from a substance.

IDEAL GAS LAW (CHAPTER 2)

When all collisions between the atoms or molecules in a gas are completely elastic, and absolutely no forces of attraction exist between those molecules or atoms, a gas is said to be in its ideal state. When this happens, all the internal energy of the gas is in the form of kinetic energy, and any change in this kinetic energy is accompanied by a change in temperature.

When we look at ideal gases, we are concerned with the following three variables: absolute pressure (P), volume (V), and absolute temperature (T). The relationship between these three variables is expressed by the ideal gas law, which is stated in the equation $PV = nRT = NkT$, where the following definitions apply:

n = number of moles
R = universal gas constant = 8.3145 J/mol K
N = number of molecules
k = Boltzmann constant = 1.38066×10^{-23} J/K =
$\quad$ 8.617385×10^{-5} eV/K

$k = R/N_A$

N_A = Avogadro's number = 6.0221 x 10^{23} /mol

KINETIC THEORY OF MATTER (CHAPTER 2)

The kinetic theory of matter states that matter is composed of a large number of small particles—either individual atoms or molecules—that are in constant motion. This theory, also known as the kinetic molecular theory of matter and the kinetic theory, helps to explain how matter behaves, especially concerning the transfer of heat between materials and the relationship between pressure, temperature, and volume properties of gases.

Several assumptions are made in the kinetic theory of matter in order to keep the theory from becoming too complex. The first assumption, as already stated, is that the matter is made up of very small particles, either atoms or molecules.

The next assumption is that the spacing between these particles determines the form that the matter takes. In a gas, for example, the separation between particles is very large, allowing enough space to prevent any forces of attraction or repulsion between the particles. In a liquid, the particles are closer together, close enough that forces of attraction come into play and confine the liquid to its container. In a solid, the particles are densely packed, creating forces of attraction that are so strong they confine the material to its shape.

A third assumption is that the particles are in constant motion. As with the spacing between particles, the freedom of movement of the particles helps to determine the form of the matter. In gases, the movement of particles is said to be random and free. In liquids, the movement is somewhat constrained by the container, and in solids the movement is completely restricted by the form.

The fourth assumption is that when atoms or molecules collide, there is an exchange of energy. The speed at which they are traveling when they collide determines the kinetic energy created by the collision.

THERMODYNAMICS (CHAPTER 5)

Thermodynamics is the study of the movement of energy, or heat, between systems and the relationship of work to this movement. The principles of thermodynamics have played a huge role in the modernization of society because many machines and devices, such as the automobile, change heat into work or work into heat. Thermodynamics is governed by four fundamental laws.

FIRST LAW OF THERMODYNAMICS (CHAPTER 6)

The first law of thermodynamics is also called the law of the conservation of energy. It states that while energy can change form and be transferred between systems, it cannot be created or destroyed. It can be converted from mechanical work to heat, from heat to light, or from chemical to heat, for example. This is equivalent to stating that the amount of energy in the universe is a constant.

SECOND LAW OF THERMODYNAMICS (CHAPTER 6)

The second law of thermodynamics addresses entropy and the direction of energy, or heat, flow in the universe. It states that heat cannot flow from a cooler object to a warmer object unless work is applied to the system. The second law also states that some energy will always be lost in the form of entropy in a thermodynamic system because entropy, defined as the amount of disorder in a closed system, is always increasing.

THIRD LAW OF THERMODYNAMICS (CHAPTER 6)

The third law states that the entropy of a system at absolute zero is a well-defined constant and uses the example of a perfect crystal to illustrate the point. As the absolute temperature of a perfect crystal approaches zero, the entropy of that perfect crystal is said to approach zero. This law provides an absolute reference point for the determination of entropy.

GLOSSARY

ABSOLUTE ZERO   A theoretical concept of the lowest possible temperature, at which molecular motion is at its lowest possible point. By definition, no temperature can be lower than absolute zero.

ADIABATIC PROCESS   A thermodynamic process during which no heat energy enters or leaves the system.

BLACK BODY   An object that is a perfect absorber and a perfect emitter of radiative energy.

CALORIE   A unit of heat measurement, equal to the amount of energy required to heat one gram of water by one degree Celsius or 4.186 Joules.

CLASSICAL THERMODYNAMICS   The study of thermodynamics through the observation of macroscopic changes in a system, without determination of the mechanisms of such changes.

CONDENSATE   An exotic state of matter that only exists at extremely low temperatures. Matter in this state exhibits behaviors of waves rather than particles and possesses the property of superconductivity.

CONDENSATION   The shift of matter from gas to liquid (not to be confused with *condensate*).

CONDUCTION   Heat transfer between two objects in physical contact, due to the collision of molecules.

CONVECTION   Heat transfer through the movement of molecules in a fluid and the collisions of those molecules.

DEFORMATION   A change of shape resulting from pressing a force onto an object, which is resisted only by matter in the solid state.

DEIONIZATION   The shift of matter from plasma to gas; molecules regain lost electrons and become neutrally charged.

DEPOSITION   The shift of matter from a gaseous state directly to a solid.

ELASTIC   A type of collision in which all of the energy of motion contained within the colliding objects is conserved, and the objects that collide bounce off of one another in the opposite direction but at equal speed of the initial motion.

ENDOTHERMIC REACTION   A reaction with a positive ΔH (change in enthalpy), meaning it absorbs heat from its surroundings.

ENTHALPY   A measurement of the energy of an object, including both the internal energy of the object and the energy required for that object to retain its shape at a specific pressure.

ENTHALPY OF CONDENSATION   Amount of energy required for one mole of a substance to shift from gaseous to liquid state.

ENTHALPY OF DEIONIZATION   Amount of energy required for one mole of a substance to shift from plasma to gaseous state.

ENTHALPY OF DEPOSITION   Amount of energy required for one mole of a substance to shift from gaseous to solid state; equal to the sum of the enthalpy of condensation and the enthalpy of freezing.

ENTHALPY OF FREEZING   Amount of energy required for one mole of a substance to shift from liquid to solid state.

ENTHALPY OF FUSION   Amount of energy required for one mole of a substance to shift from solid to liquid state.

ENTHALPY OF IONIZATION   Amount of energy required for one mole of a substance to shift from gaseous to plasma state.

ENTHALPY OF SUBLIMATION   Amount of energy required for one mole of a substance to shift from solid to gaseous state; equal to the sum of the enthalpy of fusion and the enthalpy of vaporization.

ENTHALPY OF VAPORIZATION   Amount of energy required for one mole of a substance to shift from liquid to gaseous state.

ENTROPY   A measurement of the amount of disorder in a system; a measurement of the amount of energy in a system that cannot be used to perform work.

ENVIRONMENT   In thermodynamics, all space that is not within the area defined by the system being studied.

EVAPORATION   The ability of molecules in a liquid to escape in the form of gas below the usual temperature of vaporization.

EXOTHERMIC REACTION   A reaction with a negative ΔH (change in enthalpy), meaning it releases heat into its surroundings.

FREEZING   The shift of matter from liquid to solid.

GAS   A state of matter with an undefined volume and shape, so it will expand to take on the size and form of a container.

HEAT   The lateral motion of the molecules within an object.

HEAT ENGINE   A machine for the conversion of thermal energy into mechanical or electrical energy.

HEAT PUMP CYCLE   A thermodynamic cycle in which mechanical work is used to change the temperature of the environment surrounding a system.

HEAT TRANSFER   The movement of heat energy within an object or between two objects.

IDEAL GAS   A theoretical gas made up of noninteracting point particles. The concept is similar enough to real-world gases to be used as an approximation.

INTERNAL ENERGY   The total energy contained within the molecules that make up a particular object.

IONIZATION   The shift of matter from gas to plasma; molecules lose electrons and become positively charged.

IONS   Molecules with an electric charge due to having an abnormal number of electrons. The ions that make up a plasma are positively charged, because they have lost one or more negatively charged electrons.

ISOBARIC PROCESS   A thermodynamic process during which heat energy is transferred but no change in pressure occurs.

ISOMETRIC PROCESS   A thermodynamic process in which heat energy is added to a system but the volume of that system remains constant.

ISOTHERMAL PROCESS   A thermodynamic process in which heat energy is transferred but no change in temperature occurs.

JOULE   A unit of measurement for energy or work, equal to 0.239 calories.

KINETIC ENERGY   The energy of motion.

KINETIC THEORY OF MATTER   All matter is made up of smaller particles, which are in constant motion. This molecular motion then defines the attributes of the matter.

LATENT HEAT   The energy added to an object at its phase transition point, which does not result in a change in temperature of that object.

LIQUID   A state of matter with a defined volume but an undefined shape, meaning it will take on the form of a container into which it has been poured.

MACROSCOPIC   Observable by the human eye without magnification equipment.

MELTING   The shift of matter from solid to liquid.

MPEMBA EFFECT   The observation that warmer water sometimes freezes faster than colder water.

PERFECT CRYSTAL   An arrangement of molecules in a solid that is a repetition of the same pattern without deviation or imperfection. This concept is most commonly referred to in reference to the third law of thermodynamics.

PERPETUAL MOTION   An impossible scenario in which a machine can be constructed such that it will never run down even without the addition of energy from outside the system. Perpetual motion is in violation of the first and second laws of thermodynamics.

PHASE CHANGE   The shift of matter when it changes from one type to another, for example, when liquid water freezes into ice.

PHASE TRANSITION POINT   The temperature and pressure for a specific material at which it will undergo a change of phase, such as freezing.

PLASMA   A state of matter made up of ionized particles with undefined volume and shape, and high levels of electrical conductivity.

POWER CYCLE   A thermodynamic cycle in which heat energy is transferred into a system and used to perform work.

PRESSURE   The amount of force exerted by the molecules of a fluid through collisions with the walls of a container.

RADIATION   Heat transfer due to the electromagnetic energy emitted by matter.

RADIATIVE EQUILIBRIUM   A state at which an object is emitting and absorbing equal amounts of energy through radiation.

REPRODUCIBLE   A property of good research data that means the same results will be determined by another researcher who performs the same experiments. Results that are not reproducible are not considered to be scientifically accurate.

SOLID   A state of matter with a defined shape and defined volume, which can resist deformation.

SPECIFIC HEAT   The ability of a substance to resist a change in temperature.

SPECIFIC HEAT CAPACITY   The amount of heat energy required to raise one gram of that substance by 1°C, calibrated against the calorie.

SUBLIMATION   The shift of matter from solid directly to gas.

SUPERCONDUCTIVITY   A property of condensates that causes them to have an electrical resistance of zero.

SUPERCOOLING   The chilling of water to below its freezing temperature without triggering the formation of ice.

SYSTEM   In thermodynamics, a collection of objects within a defined space that are related to one another by a set of observable properties. A system was originally defined by Carnot as a "working substance under study."

TEMPERATURE   A measurement of the amount of heat energy within an object, relative to a set scale.

THERMAL   Dealing with or relating to heat and/or temperature.

THERMAL CONTACT   Any situation where heat energy could be transferred from one object to another.

THERMAL CONDUCTIVITY   The rate of heat transfer through conduction for a particular material.

THERMAL EQUILIBRIUM   The state of two objects when they contain equal amounts of heat energy and measure at the same temperature. While energy is still being passed between them, there is no net heat flow.

THERMODYNAMICS The study of the transfer of heat energy and the work that can be performed by this transfer; a science that relates heat transfer to the forces acting on objects.

THERMODYNAMIC CYCLE A multistep thermodynamic process that eventually returns to its original state.

THERMODYNAMIC EQUILIBRIUM A state at which the objects within a system are equal to one another in all properties, including thermal, chemical, mechanical, and radiative states.

THERMODYNAMIC PROCESSES Reactions that result in the transfer of heat energy and/or a change of phase in the system of interest.

THERMODYNAMIC WORK The ability of a process to transfer heat energy within, into, or out of a system, as opposed to mechanical work, which causes a displacement of an object.

TRIPLE POINT The temperature and pressure at which the solid, liquid, and gaseous state of a particular material can coexist.

VAPORIZATION The shift of matter from liquid to gas.

WORK The action of a force that causes a displacement of an object; also referred to as *mechanical work*.

FURTHER RESOURCES

Print and Internet

Brain, M. HowStuffWorks. Brain, M. "How Thermoses (Vacuum Flasks) Work." Available online. URL: http://home.howstuffworks.com/thermos.htm. Accessed August 1, 2011. This article discusses the mechanisms behind the workings of a simple thermos-style flask, exemplifying the realities of heat transfer and the insulating powers of a vacuum. The article also discusses the effects of the vacuum of space on keeping the environment at a suitable temperature for space travel.

Chemtutor LLC. "States of Matter." Available online. URL: http://www.chemtutor.com/sta.htm#wok. Accessed August 1, 2011. A Web site article that briefly details some basic physics facts with examples. This particular section focuses on explaining how phase changes work in water, using a graph as a visual guide to the process.

Flowers, Charles. *Instability Rules: The Ten Most Amazing Ideas of Modern Science*. New York: John Wiley & Sons, Inc., 2002. This publication presents an interesting and easily understood description of ten important principles of modern science and their impact.

Huetinck, L., and S. Adams. *CliffsQuickReview Physics*. New York: John Wiley & Sons, 2001. From the well-known publisher of the literature study aid CliffNotes, *CliffNotesQuickReview Physics* is a brief guide to the core concepts of basic physics. Chapter 3 covers the topics of heat and thermodynamics.

Nahle, Nasif. "Heat Transfer, Conduction, Convection and Radiation." *Biology Cabinet*. Available online. URL: http://www.biocab.org/Heat_Transfer.html. Accessed August 1, 2011. An article by a college professor that links the processes of heat transfer to biological and ecological systems. While many of the written explanations are out of the reach of a basic physics student, the images and graphs on the page are very clear explanations of how heat transfer affects the world around us.

NASA Glenn Research Center. "What is Thermodynamics?" Available online. URL: http://www.grc.nasa.gov/WWW/K-12/airplane/thermo.html. Accessed August 1, 2011. These articles by NASA provide good visual interpretations of the laws of thermodynamics in action. The focus is on applying the science of thermodynamics to the reality of airborne vehicle propulsion, but the applications given can work for almost any purpose.

Steiger, Frank. "The Second Law of Thermodynamics, Evolution, and Probability." *The TalkOrigins Archive.* Available online. URL: http://www.talkorigins.org/faqs/thermo/probability.html. Accessed August 1, 2011. An article that connects the second law of thermodynamics, about increasing entropy, to the concept of biological life. The focus of the discussion is on why the second law does not explicitly or implicitly prohibit biological change processes such as evolution.

Web Sites

HyperPhysics. Available online. URL: hyperphysics.phy-astr.gsu.edu/hbase/hph.html. Accessed August 1, 2011. A Web site, maintained by the Department of Physics and Astronomy at Georgia State University, that covers a multitude of physics topics in a straightforward and easy-to-understand manner.

NASA: For Students. Available online. URL: www.nasa.gov/audience/forstudents/index.html. Accessed August 1, 2011. A Web site that contains a great deal of information, including a description of the electromagnetic spectrum, geared toward students.

National Radio Astronomy Observatory. Available online. URL: www.nrao.edu. Accessed August 1, 2011. A Web site that describes and explains radio astronomy.

Nuffield Foundation. "Molecules in Motion. Practical Physics." Available online. URL: http://www.practicalphysics.org/go/Topic_4.html;jsessionid=aUtZhAldL-T-?topic_id=4>. Accessed August 1, 2011. *Practical Physics* is a section that includes excellent resources for students. The articles cover such topics as expansion due to temperature changes, phase changes, absolute zero, and kinetic theory.

Public Broadcasting System. "Nova: Absolute Zero." Available online. URL: http://www.pbs.org/wgbh/nova/zero/. Accessed August 1, 2011. A NOVA Web site devoted to temperature. Although the main focus is on absolute zero and the history of refrigeration, there are also interactive guides on temperature scales and the states of matter.

Virtual Laboratory: Thermodynamic Equilibrium. Available online. URL: http://jersey.uoregon.edu/vlab/Thermodynamics/. Accessed

August 1, 2011. This Web page prepared by the University of Oregon's Department of Physics provides a series of interactive applets that allow the user to explore the properties of thermodynamic equilibrium. While the graphics and the experiments are extremely simple, there seems no better way to understand this concept than by controlling it yourself.